BUILDING TO LAST

This book is dedicated with love to our grandchildren:
Christopher Jack, born May 19 1993
and
Thomas Colin, born July 4 1995
and their generation.

BUILDING TO LAST
THE CHALLENGE FOR BUSINESS LEADERS

Colin Hutchinson

earthscan
from Routledge

First published by Earthscan in the UK and USA in 1997

For a full list of publications please contact:
Earthscan
2 Park Square, Milton Park, Abingdon, Oxfordshire OX14 4RN
Simultaneously published in the USA and Canada by Earthscan
711 Third Avenue, New York, NY 10017

First issued in paperback 2016

Earthscan is an imprint of the Taylor & Francis Group, an informa business

A catalogue record for this book is available from the British Library

ISBN 13: 978-1-138-96519-5 (pbk)
ISBN 13: 978-1-85383-431-8 (hbk)
Typesetting and page design by Gary Inwood Studios

Cover design by Andrew Corbett

CONTENTS

LIST OF FIGURES

LIST OF TABLES

LIST OF ABBREVIATIONS

ACBE	Advisory Committee on Business and the Environment
Agenda 21	Agenda agreed at UNCED for global sustainable development in twenty-first century
AMED	Association for Management Education and Development
AOL	America Online
B&Q	British DIY retail store
BCSD	Business Charter for Sustainable Development (prepared by ICC)
BiC	Business in the Community
BiE	Business in the Environment
BOF	British Organic Farmers
BRE	Building Research Establishment
BREEMAN	Building Research Environmental Assessment Method
BSE	Bovine Spongiform Encephalopathy (mad cow disease)
BS5750	British Standard relating to Quality Management
BS7750	British Standard relating to Environmental Management
BT	British Telecommunications plc
CAETS	Council of Academies of Engineering and Technological Sciences
CCI	Chandpur Cottage Industries
CCPA	Canadian Chemical Producers' Association
CEO	Chief Executive Officer
CER	Corporate Environmental Report
CFCs	Chlorofluorocarbons
CHP	Combined Heat and Power
CIA	Chemical Industries Association
CJD	Creutzfeldt–Jakob Disease
CLGI	Centre for Large Group Interventions
CMA	Chemical Manufacturers Association
CO_2	Carbon Dioxide
DIY	Do It Yourself
EA	Environmental Activists, especially British Environmental Activists
EIP	Eco-Industrial Parks
EIRIS	Ethical Investment Research Services
EMAS	Eco-Management and Audit System
EMS	Environmental Management Systems
ENDS	Environmental Data Services
EPA	Environmental Protection Agency (USA)
ERRA	European Recovery and Recycling Association
ESI	Ethics, enlightened Self-interest, Innovation (author's own acronym)
FAO	Food and Agriculture Organisation (UN Agency)
FOE	Friends of the Earth
FSC	Forest Stewardship Council
GAP	Global Action Plan for the Earth
GATT	General Agreement on Tariffs and Trade

GCs	Green Consumers, especially British Green Consumers
GEMI	Global Environmental Management Initiative
GEN	Global Eco-village Network
GNP	Gross National Product
GWP	Global Warming Potential
ICC	International Chamber of Commerce
IE	Industrial Ecology
IEM	Institute of Environmental Management
IHEI	International Hotels Environment Initiative
IGD	Institute of Grocery Distribution
IPCC	International Panel on Climate Change
ISDRN	International Sustainable Development Research Network
ISEW	Index of Sustainable Economic Welfare
ISO	International Standards Organisation
IT	Information Technology
ITDG	Intermediate Technology Development Group (sometimes abbreviated to IT)
ITTO	International Timber Trade Organisation
IUCN	The World Conservation Union
LGMB	Local Government Management Board
LA 21	Local Agenda 21 (see Agenda 21 above)
LCA	Life Cycle Assessment
LETS	Local Exchange Trading Systems
MIT	Massachusetts Institute of Technology
MORI	Market and Opinion Research International
MSC	Marine Stewardship Council
NEF	New Economics Foundation
NGO	Non Governmental Organisation
ODP	Ozone Depleting Potential
OECD	Organisation for Economic Cooperation and Development
OGA	Organic Growers Association
OPM	Office for Public Management
P&G	Procter & Gamble
PCR	Post-Consumer Recycled Content
PEP	Personal Equity Plan
PERI	Public Environmental Reporting Initiative
PR	Public Relations
QUANGO	Quasi-Autonomous Non-Governmental Organisation
RMI	Rocky Mountain Institute
RSA	Royal Society for the encouragement of the Arts, Sciences and Manufacturing
RSPB	Royal Society for the Protection of Birds
RX	Rank Xerox
SAFE	Sustainable Agriculture, Food and Environment Alliance
SARA	United States 'Super Fund Act' (see p73)
SCEP	Study of Critical Environmental Problems
SDN	Sustainable Development Network

SEA	Single European Act
SEAs	Sustainable Energy Alternatives
SHE	Safety, Health and Environment (Department) – sometimes HSE
SMAS	Sustainable Management and Audit Scheme
SME	Small or medium-sized enterprise
SQM	Sustainable Quality Management
TQEM	Total Quality Environmental Management
TQM	Total Quality Management
UN	United Nations
UNCED	United Nations Conference on Environment and Development (Brazil 1992)
UNEP	United Nations Environmental Programme
UNESCO	United Nations Educational, Scientific and Cultural Organisation
VAT	Value Added Tax
WBA	World Business Academy
WBCSD	World Business Council for Sustainable Development
WCS	World Conservation Strategy
WI	Worldwatch Institute
WICE	World Industry Council for the Environment (now part of WBCSD)
WICEM	World Industry Conference on Environmental Management
WWF	World Wide Fund for Nature (previously World Wildlife Fund)

ACKNOWLEDGEMENTS

Many people have helped me plan and write this book. Tom Haddon invited me to speak at the annual conference of the Strategic Planning Conference 'Creating the Strategic Enterprise' in November 1995. This stimulated the research which formed the basis of my talk. The material collected warranted an article which was published in *Long Range Planning* in February 1996 under the title of 'Integrating Environment Policy with Business Strategy'. Bernard Taylor's encouragement proved very helpful. My colleagues in the Association for Management Education and Development (AMED)'s Sustainable Development Network have helped me refine my ideas and add insights. This support encouraged me to develop the ideas, write a book and seek a publisher.

Jonathan Sinclair Wilson of Earthscan liked the idea of the book and has supported me throughout its preparation. His challenging and helpful observations have improved the book in many ways and given encouragement.

Comments on the early chapters were received from Jacquie Burgess, Nick Hutchinson, Peter Martin, Jonathan Sinclair Wilson and Coro Strandberg. Their feedback persuaded me to make major changes to Part 1 which sets the scene for the rest of the book. Tony Dearsley provided useful comments on Chapters 5 and 6. It was not always possible to do justice to the many ideas that were offered but several have been incorporated in the final version.

Specific help on particular sections in the book and factual information on various topics was provided by Arthur Brian, Michael Boddington, Charter 88, Laura Cooper, Pat Dade, Richard Evans, Simon Guy, Tony Hams, Kim Jonas, Francis Kinsman, Martin Leith, Suzanne Pollack, Philip Prideaux, Robert Rasmussen, Trewin Restorick, the Royal Academy of Engineering, the Soil Association, Bernadette Thorpe, the Water Services Association, Richard Welford, Brian Whitaker, Martin White and the World Wide Fund for Nature.

A special word of thanks must go to Jim Hopwood who read the first seven chapters and provided many useful comments. For Chapter 5 help from the companies whose work I chose to describe was needed and I am indebted to David Hammond and Peter Hindle of Procter & Gamble, Irina Maslennikova and Pierre van Coppernolle of Rank Xerox, Per Grunewald and Henrik Troberg of Electrolux, Paul Monaghan of The Co-operative Bank, Alan Knight of B&Q, and Steve Tomlin and Nick Hackett of Reclamation Services. Many companies, other organisations, and sources of published material are acknowledged, as appropriate throughout the book.

Peter Hindle and Philip Sadler read the first draft of the whole book and their insights, observations and comments added ideas which strengthened many parts of the book. Robert Rasmussen, Ernie Lowe, and Erich Schwarz provided invaluable help on Kalundborg, Industrial Ecology, and Eco-Industrial Parks which are described in Chapter 7.

Fiona Murray proof read the final version and enabled me to clarify wording and improve the flow of ideas throughout the whole book. My family and Pat, my wife, as always, were extremely tolerant of the many hours spent working on the book and coping with me during the frustrations at various stages in the process.

To all these people I am indebted but the final text remains my responsibility.

Colin Hutchinson
Oxted, Surrey
10 December 1996.

FOREWORD

This stimulating book comes out at a time of significant change in the environmental field. The first change, which is the main theme of the book, concerns the growing awareness that the key issue is not just environmental improvement but, more broadly, sustainability. It will no longer be sufficient for industrial companies to improve their environmental performance, for example by reducing emissions, recycling waste and optimising the use of their raw materials; they will have to integrate sustainability into the core of their business strategy. The second change is that the emphasis amongst most environmental non-governmental organisations (NGOs) is switching from campaigning against 'business' to working in partnership with it. It has become a cliché to suggest that companies are not the problem but the solution in the environmental field. The reason is that, in a free enterprise system, entrepreneurial companies will develop the technology to eliminate problems. With governments and NGOs promoting sustainability (embracing economic and social aspects as well as environmental ones) in a positive manner, and with companies developing solutions, progress is likely to be faster and more effective. The third change is that there is a growing realisation that the greatest damage to the planet which we inhabit stems from the size and behaviour of the human population. For example, in most of the developed world, air pollution is far more dependent on the emissions of vehicles than the emissions of industry. Since it is so hard to change the behaviour of people, once again it is likely to be the ingenuity of companies which will ameliorate the problem.

Of course, some large companies still see environmental issues as a chore or a fad, whilst for most small companies they remain largely an unwelcome distraction. However, a growing number of companies of all sizes accept that reconciling profitability and sustainability (the theme of the Prince of Wales's Business and the Environment Programme) is the critical issue. The most far-sighted understand that even if they cannot yet grasp fully what sustainability will entail, at least it will provide an opportunity for gaining competitive advantage.

For those who are not yet convinced that radical change is required, this book should help them to change their minds. The early chapters lucidly explain the causes for concern, the need for a radical approach to dealing with local and global problems and what is meant by sustainability. But business people have to be practical to succeed, and Part 2 of the book gives a wide range of examples of initiatives by businesses and other organisations, which are proving successful in moving towards a more sustainable society. Part 3 examines in more detail the various approaches and systems which have been adopted in the pursuit of sustainability.

Although this will prove to be a valuable textbook for environmental causes, it is likely to prove even more useful for business people, to consultants advising businesses and other organisations, to environmentalists working in partnership with business, and, finally, to governments and their concerned but often ill-informed citizens. I commend it to the wide audience it deserves.

John G Spiers
Managing Director, Norsk Hydro (UK) Ltd
February 1997

INTRODUCTION

Creating a successful company with excellent performance and the ability to maintain its success has never been more challenging. The rate of change, complexity of regulations and speed of communication all contribute to this challenge. Despite this the desire to be associated with a company that has lasting success remains strong. Helpful guidelines towards such success are now available as a result of several research studies.

The successful habits of visionary companies have been the subject of extensive research that will dispel many myths.[1] Successful companies with a reputation for being visionary are compared with other companies of comparable size in the same sector. The researchers explored the reasons for the outstanding performance of the visionary companies with some surprising results. They often start slowly but make steady progress. They do not have high profile, charismatic leaders. Their primary goal is not to maximise profits but to pursue several objectives, one of which is to make money. The core values are often both enlightened and humanistic. Their employees enjoy working in these companies because they fit in well and contribute a lot, or they dislike it and leave. There is little middle ground. Rather than concentrating on beating the competition these companies focus on constant improvement of their own standards. They avoid the 'tyranny of OR' by adopting the 'genius of AND'. This means that they reject dichotomies such as stability or progress, environment or profits, conservative or audacious goals by accepting apparently conflicting objectives and pursuing both. An important ingredient for success is corporate ideology, the combination of core values and statement of purpose, which is the primary guiding principle and is adhered to even in adversity.

Another important criterion for success is organisation culture. Many managers believe that the most successful companies have a strong organisation culture and that this is the essential ingredient for lasting success. Others argue that a good fit with the organisation's mission and the organisation's culture is the vital requirement. A careful study of organisation culture and business results has exploded both these beliefs as myths.[2] The most successful companies are those which have adaptable organisation cultures. This is not surprising in an era of unprecedented change but changing an organisation's culture is never easy. It requires skill and perseverance but it can be done as several companies have shown.

A third area of interest relates to customer focus. Those who believe that the market is unchanging and that the customer will remain loyal may be in for a shock. [3] There is a pressing need to be aware of changing customer awareness, values and needs. This is not new and has been the subject of many best selling business books over several decades.[4]

These studies provide a starting point for organisations wishing to achieve excellent performance. The added dimensions that are now being included are ecology and sustainability. These now need to be integrated with the best ideas of visionary companies: sound corporate ideology, adaptable organisation cultures and strong customer focus. Organisations are most likely to achieve enduring performance by understanding how to build an organisation that will last.

ECOLOGICAL, SOCIAL AND ECONOMIC SUSTAINABILITY

There is widespread recognition that ecological and social problems affect every country, every business and everybody. Solutions that do not destroy economic stability are needed badly and increasingly widely sought. The actions of many businesses demonstrate that sound environmental management makes business sense. Organisations like 3M, Norsk Hydro and Dow Chemicals have pioneered environmental management and publish their achievements. These companies have not only reduced their environmental impact but have also achieved a good return on their investments. They are not alone. Three out of four of the UK's largest companies have environmental policies, over 2000 companies worldwide have signed the Business Charter for Sustainable Development. A few, including British Telecommunications, are planning to carry out a social audit.[5]

Sometimes it is hard for businesses to declare what they have done and how they have reduced some of their adverse effects on the natural world. When an example is noted by the media or an active environmental group, they are often quick to decry the success and point out what still needs to be done. Environmental managers are often disappointed by the lack of interest in their company's environmental reports. The disappointment can be disheartening because a lot of effort has gone into the initiatives and producing the report. The combined effect of being criticised or ignored, or both, discourages many businesses from trying to publicise what they are doing about environmental management. Their initiatives make business sense so they just get on with it. There is much more being done by business than is generally known. One of the aims of this book is to describe some good practices in order that they become better known.

There is, of course, the darker side too. Several large and powerful companies actively campaign to block legislation and fiscal measures that may help the environment but which may add to their costs in the short term. This is not explored in this book as others have covered the topic in some detail.[6]

Some companies help governments to develop appropriate regulations. They do this because they know that badly drafted legislation and ill-conceived fiscal measures fail to achieve the intended results. A cooperative approach can help to bring in measures that make a positive contribution and everyone benefits if there is commitment to abide by the agreed regulations.

Many companies prefer voluntary codes to regulation. A wide range of charters, codes of practice and industry guidelines is available. One of the best known is the Business Charter for Sustainable Development but even the International Chamber of Commerce (ICC), who created the Charter, would like its use to be more widespread and effective. The ICC cannot penalise those who ignore the charter, sign it but do nothing, or use it ineffectively. A few companies take it seriously and report progress regularly. There are others that do little, many that ignore it, and millions of businesses that have never heard of the Charter or any other voluntary code.

One of the difficulties is that environmental management is widely assumed to be a costly matter. The emphasis has been on cutting emissions and reducing wastes for disposal. More attention could be given, especially by smaller business, to three important opportunities. First they could appreciate that investment in reducing emissions and waste at source often gives a quick return. Second, retaining the right to be an approved supplier to a large business increasingly requires high standards of environmental management. Third, new business opportunities are emerging from goods and services which enable others to cut emissions, reduce wastes, monitor performance and make progress towards sustainability.

THIS BOOK

This book sets the scene by describing our changing world, then looks at three approaches being taken by companies, professional bodies and communities to face the new challenge. It concludes with practical ideas for managing the transition, measuring progress and making change happen. The objectives of this book are to:

1   summarise how ideas about ecological, social and economic sustainability have evolved;
2   illustrate ways in which some people are changing their behaviour;
3   describe how some companies are building their businesses to last;
4   indicate how professional bodies and communities are making their contribution; and
5   suggest some processes that can make it happen.

The three parts of the book start with an outline of each chapter. The primary aim is to provide a practical book for senior managers. Examples show how ideas are being put into practice by successful companies. While the book is mainly for business managers many of the ideas are relevant for the public and voluntary sectors. Those who work with businesses or who work to help bring about change in the community will also find ideas they can apply in their work.

The section summaries should help the reader to follow the flow of ideas. The notes at the end of the book provide references for those who wish to carry out further research on selected topics. Tables and figures add more detail to complement the text.

Three appendices provide additional supporting material. Appendix 1 describes why and how some companies have taken significant action. It illustrates that each situation is unique and different circumstances provide the trigger for new initiatives. Appendix 2 is a brief summary of a survey of small firms and the environment, indicating their special situation and their needs. Appendix 3 offers a summary that illustrates what the transition from the old thinking to new thinking may mean.

A central proposition throughout the book is that ethics and enlightened self interests are at the heart of lasting solutions. Ethics, in this sense, includes responsibility towards present *and* future generations, to economics *and* ecology, and obligations towards the individual *and* society. Organisations that are built to last are likely to pursue multiple objectives and accept the wider responsibilities of social and ecological performance alongside their economic objectives. This requires the courage to accept the challenge and the wisdom to move forward in partnership with others.

PART I
OUR CHANGING WORLD

For some 30 years scientists have been refining their perceptions of the threats to the Earth. Some uncertainties remain but there is now a strong global consensus about the need for action to protect the Earth's life support systems. Everyone has a part to play.

The chapters in this section describe different aspects of our changing world:

- the gradual emergence of a common global purpose;
- ways in which people are experimenting with changes in their own lives;
- the relevance of market trends and new mindsets for business; and
- what sustainable development and sustainable enterprise mean.

Chapter 1 describes some of the major landmarks that contribute to a broad set of ideas suggesting that a global common purpose is emerging. The origins of these ideas go back a very long way but for practical purposes the significant developments have occurred in the last 50 years. The events which are particularly noteworthy take three forms. Firstly, major scientific and social studies which have led to the publication of authoritative and influential reports. They draw attention to global problems and suggest possible solutions. Secondly, there has been much pioneering work carried out 'in the field' by people who have gone on to establish environmental groups. These organisations have done so much to add insights to human understanding of the living world. Thirdly, a few major events vividly demonstrate how the natural world can be and is being damaged by human activity.

The brief chronological summary is followed by a description of some of the initiatives being taken by business leaders and the guidelines they are proposing for general application by enterprise of all kinds. The combination of landmark publications, insights from practical work in the field, and a selection of damaging events provides the basis for a summary of the major threats to the Earth. The chapter concludes by drawing on proposals put forward by Al Gore, before he became the US Vice President, for a common purpose and global plan. Business leadership has a major part to play in implementing this plan.

Chapter 2 looks at ways in which Gross National Product (GNP) as a measure of progress in society, fails to take account of several factors that contribute to quality of life. This is giving rise to other measurements of progress, such as the Index of Sustainable Economic Welfare (ISEW). Quite apart from the measurement methods used there are strong signs that traditional beliefs are giving way to emerging ideas and some of the areas in which this is happening are noted.

Dissatisfaction with some traditional aspects of society leads some people to try out new ideas. These include investing in companies that are screened for their ethical and environmental performance, and a steady shift away from fossil fuel energy towards renewable forms of energy. Householder initiatives include seeking new forms of credit, taking part in local exchange trading systems and making lifestyle changes at home. The latter include better ways to deal with household waste, using energy, transport and water more efficiently, modifying buying habits and diets and making use of complementary medicine. The last section of this chapter describes the campaign for a new constitution in the UK with improved democracy and more social justice. This initiative now links with many voluntary organisations and charities, working in different fields, through the Real World Coalition.

Chapter 3 looks at data that indicates that markets are changing and the mindsets of managers are beginning to change as well. Surveys carried by organisations such as ENTEC and MORI help to describe market trends that have considerable relevance for businesses of every kind. The RSA's report on Tomorrow's Company indicates that British business lacks competitiveness and fails to give adequate attention to customers, suppliers and employees. They recommend an inclusive approach which acknowledges the importance of all stakeholders.

Improved product labelling is gaining ground, with cooperation from and implications for businesses of all kinds. These labels help customers make informed buying decisions. There are also social surveys which aid understanding of the ways in which people in society can be classified and their behaviour studied. This can be very helpful for designing products, marketing and advertising, all of which helps business to meet target markets in cost-effective ways.

While some businesses are taking the environment very seriously there are others that are not. Some of the reasons for this are explored and a variety of mindsets are described to illustrate ways in which enlightened development can take place. The chapter concludes with some thoughts about a personal agenda for action.

Chapter 4 looks at the question of sustainable enterprise. There is a clear role for government to establish the framework conditions which enable sustainable enterprise to emerge. The specific contribution of government is outlined. It is also acknowledged that no single country or business can become sustainable alone, but it is emphasised that everyone can play a part.

Any form of development is a process rather than a destination. The destination for sustainable development is to create societies that are economically viable, socially desirable and ecologically sustainable. A vision of a society like this is described to illustrate that it is not difficult to develop visions of this kind. If such a vision can be widely understood it can provide the focus for change and development for individuals and organisations in every sector of our society.

One of the most practical ideas for implementing sustainable development has come from The Natural Step in Sweden. Their work is summarised as are their guidelines for business. When it comes to implementing action a few of the priority areas are identified, such as energy and transport.

Business leadership in building businesses which last will contribute to the creation of sustainable societies, but the reverse is also true. As initiatives from government and within communities apply the ideas of sustainable development they help create the conditions in which sustainable enterprise can flourish.

I AN EMERGING
GLOBAL COMMON PURPOSE

The evolution of environmental awareness is gradually producing a shared understanding worldwide of the need to live within the Earth's capability to sustain life. The scientific bases of the problems and the solutions are being refined and better understood. More people are learning what this means for them at home and at work. There is now wide recognition that robust solutions depend on sound ecology, viable economics and social justice.

This chapter introduces the subject by exploring four themes:

1) the evolution of environmental awareness;
2) ways in which business leaders are becoming involved;
3) a brief summary of the threats to Earth and society; and
4) a common purpose and global plan.

THE EVOLUTION OF ENVIRONMENTAL AWARENESS[1]

Ideas for sustainable development arose out of concern for the natural environment and the extremes of poverty in parts of Africa, Asia and Latin America.

The origins of enviornmental concerns can be tracked back a very long way to. Plato in 400 BC. He was worried about deforestation and soil erosion and commented on the number of people a given area of land could support. Pliny in AD 1 described how poor animal husbandry was threatening crops. In AD 1000 it was noted how shipbuilding was destroying the Mediterranean's coastal forests, especially around Venice and Genoa.

In 1660 John Evelyn, a naturalist and one of the founders of the Royal Society, described how the burning of coal was creating a 'hellish and dismal cloud [in the City of London] so that it resembled the suburbs of hell'. In the USA William Penn in 1681 required Pennsylvania to 'leave an acre of trees for every five acres cleared'.

In 1827 John James Audubon, a naturalist and artist, published the first part of his *Birds of North America* with widely acclaimed colour plates. In the 1860s the first groups protecting the environment and wild birds were formed in the UK and the first air pollution law was passed in 1863. In the USA George Perkins Marsh helped to establish The Smithsonian Institution in 1846 and published *Man and Nature* in 1864. The National Audubon Society was formed in 1886 for the study and protection of birds. Yosemite National Park and the Sierra Club were formed in the early 1890s to protect the natural environment.

In the 1930s the USA experienced one of the worst environmental disasters on record. As a result of ill-advised agricultural practices some 500,000 square miles (1.29 million square kilometeres) of pasture was turned to dust and over 200 dust storms hit the Great Plains of the USA. Sixteen States were affected and the USA was obliged to import wheat. The Great Plains Committee made many recommendations for future safeguards.

By 1940 international treaties had been signed on nature and natural resources and toxic substances (including radiation). By 1950 treaties on protection of marine fisheries and animals were agreed. The late 1940s and early 1950s saw the emergence of the first Clean Air Acts in the UK and growth of environmental groups. The Royal Society for the Protection of Birds (RSPB) was one of these organisations. Only women members were allowed when it formed in 1889, with 5000 members in its early years but men were soon admitted. The RSPB has grown rapidly and now has nearly 1,000,000 members and employs over 700 people.

In 1946 Peter Scott, an accomplished wildfowl artist, founded the Severn Wildfowl Trust (later the Wildfowl and Wetlands Trust).[2] Other national and international organisations were formed around this time including the International Union for the Protection of Nature, a UN agency, later to become the International Union for the Conservation of Nature and Natural Resources (IUCN). The IUCN was formed in Switzerland after a prolonged gestation. Its unique Constitution was signed by 18 nations, 7 international bodies and 107 independent bodies. Its goals were ambitious but omitted any reference to population growth and its funding was inadequate.

During the 1950s, 19 international treaties, conventions and protocols were added to the 9 agreements which had been signed between 1911 and 1950. These treaties covered pollution, fisheries, natural resources, toxic substances, animals, plants and birds. A further 80 international agreements relating to the natural environment were signed between 1960 and 1985.

The 1960s

The 1960s heralded increased activity and a growing consensus. In 1961 the World Wide Fund for Nature (WWF - originally the World Wildlife Fund) was formed as a result of an initiative taken by Julian Huxley (Director-General of UNESCO), Max Nicholson (Director-General of the UK's Nature Conservancy) and Peter Scott. In 1961 Rachel Carson, a biologist, published *Silent Spring*, which became a best seller worldwide.[3] It vividly drew attention to the damage caused to wildlife by pesticides and suggested that human life itself could be threatened.

The National Research Council of the US National Academy of Sciences set up the Committee on Resources and Man and published their recommendations in 1969.[4] The result of their work was summarised as follows:

Since resources are finite, then, as population increases, the ratio of resources to man must eventually fall to an unacceptable level. This is the crux of the Malthusian dilemma, often evaded but never invalidated...The inescapable central conclusion is that both population control and better resource management are mandatory and should be effected with as little delay as possible.

The Massachusetts Institute of Technology (MIT) sponsored the Study of Critical Environmental Problems (SCEP).[5] Among the problems they considered were climate and ecological effects of man's activities, monitoring methods and availability of data, the implications of remedial action, industrial products and pollutants, domestic and agricultural wastes and energy products. A great many recommendations were made, concentrating on the need for scientific information and improved monitoring of environmental effects.

In 1967 the *Torrey Canyon* oil tanker ran aground off the west coast of England spilling 117,000 tons of crude oil. The use of untested chemicals added to the damage to birds and marine life and the total clean-up bill came to £6 million. This accident led to the formation of the Royal Commission on Environmental Pollution.

In 1968 the Club of Rome was formed at the instigation of Aurelio Peccei, an Italian industrialist. He called a meeting in Rome of 30 eminent people who were close to national and international decision makers. They were concerned about the apparent inability of world leaders to foresee the consequences of substantial material growth, the implications of unprecedented affluence and their reluctance to consider quality of life. The Club has always been a select group of 50 to 100 people with no political ambitions.[6] Their approach is based on three overlapping concepts, namely: global issues, longer term perspectives and a deeper understanding of the inter-relationship of contemporary problems. This includes political, economic, social, cultural, psychological, technological and environmental matters.

Friends of the Earth was formed in San Francisco, USA in 1969 by David Brower, former Director-General of the Sierra Club. He had been criticised for giving too much attention to campaigns and not enough to administration. The dispute was resolved when he left so that he could give more time to campaigns in a new organisation.

The 1970s

Earth Day was celebrated by hundreds of thousands of Americans in 1970. The USA Environmental Protection Agency (EPA) was formed in 1970 with a staff of 8000 and a budget of US$450 million. By 1981 this had grown to 13,000 staff and a budget of US$1.35 billion.

The UK arm of Friends of the Earth was formed in 1971 to campaign on environmental issues. Greenpeace was formed in 1977 to challenge the whaling nations and to campaign against pollution.

Green Political Parties were formed in New Zealand in 1972, Britain in 1973, France in 1974, Belgium and West Germany in 1978, and in Switzerland, Luxembourg, Finland, Sweden, Ireland, The Netherlands and Italy during the next five years.

The Club of Rome had commissioned various studies on the 'predicament of mankind', including the *Limits to Growth*, published in 1972.[7] This report concluded that the planet's carrying capacity would be exceeded sometime within the next 100 years and would result in a massive ecological catastrophe, famine and wars. It gained wide publicity but was opposed on the grounds that it exaggerated the lack of resources, under-estimated the potential for technological solutions and advocated cessation of economic growth.

The first UN Conference on the Human Environment took place in Stockholm in 1972 with the underlying fear of the developing countries that measures to protect the environment would probably go against their interests. In tackling this issue in the preparatory meetings, held the year before the conference, the links between environment and development began to emerge. The aim was to find out how development could take place while the worst pitfalls of environmental damage were avoided. One result of this work was to broaden the concept of environment to include soil erosion, the spread of deserts, water supply, and human settlements.

Control of pollutants, resource management and education were already on the agenda for the main conference. The conference organisers commissioned a report to brief delegates for their deliberations, entitled, *Only One Earth*.[8]

Following the Stockholm conference the United Nations Environment Programme (UNEP) was formed with offices in Nairobi, Kenya and with Maurice Strong as Secretary General. He had been Secretary General for the Stockholm conference and prior to that was a Canadian industrialist.

Other landmark publications at this time include *The Blueprint for Survival*[9] and *Small is Beautiful: a study of economics as if people mattered*[10] by E F Schumacher. Schumacher died four years after his famous book was published but he inspired the formation of the Schumacher College, the Schumacher Society, the Intermediate Technology Development Group (ITDG) and the New Economics Foundation all of which continue to develop their own specific contribution to environmental solutions.[11]

Chipko Andalan, the environmental movement in India, was formed in 1973 owing its origins to forest-based cottage industries in the foothills of the Himalayas which were concerned about tree felling leading to soil erosion and depletion of water supplies. In 1973–74 groups of local women, who saw their livelihood threatened, banded together to stop tree felling. The method they chose was literally to hug the trees, thereby preventing them from being felled until they were physically removed. This form of protest proved successful and gained wide publicity which stimulated the formation of thousands of environmental groups.

Amory Lovins and his wife, Hunter, founded the Rocky Mountain Institute (RMI) and published *World Energy Strategy* in 1975 and *Soft Energy Paths* in 1977.[12] The latter broke new ground by presenting a well-reasoned case showing how renewable energy sources could be developed to replace fossil fuels. Their predictions of US energy demand over the period 1975–90 proved significantly more accurate than the projections of both government and industry while actual demand was even less than the RMI estimates.

In 1976 the Barcelona Convention for the Protection of the Mediterranean Sea banned the dumping of mercury, cadmium, persistent plastics, DDT, crude oil and hydrocarbons and was signed by 17 countries around the Mediterranean.

In 1976 Love Canal near New York, USA became a symbol of the dangers of building houses and schools on a site previously used for dumping chemical wastes. Following unusually heavy rain, chemicals seeped into homes and the children's playground. The homes of 263 families had to be purchased and the families had to be relocated; an additional 1000 were advised to move and US$27 million was spent on temporary housing and containing the problems.

In 1979 the nuclear power station at Three Mile Island, Harrisburg, Pennsylvania was put out of action by an accident. This led to strong opposition to nuclear power development in the USA. Since then there has been a declining trend worldwide in the number of new nuclear reactor constructions started each year with none in 1995.[13]

UN Conferences have been held on many topics such as Human Settlements,[14] Law of the Sea, Population, Role of Women and they continue on a regular basis.

The 1980s

The Global 2000 Report to the President of the USA was submitted to President Carter.[15] It concluded that by the year 2000 world population would have grown by half, the gap between rich and poor would have widened, resources would be more scarce – especially land and water, life-supporting ecosystems would be reduced, prices of vital resources would be higher and the world would be more vulnerable to natural disasters and disruptions caused by humans.

Willy Brandt, former Chancellor of West Germany, chaired the Independent Commission on International Development from 1977 to 1983. The Commission published two books[16] during this period which contributed to world understanding of the tensions between industrialised and developing countries. Freedom from dependence, oppression and hunger would only be achieved by a more equitable approach worldwide and social justice.

The World Conservation Strategy (WCS) was launched in 1980 with funding from UNEP, heralding a more strategic approach.[17] The idea was put forward in 1966 but work on the strategy document was not started until 1977. Leadership was provided by the International Union for the Conservation of Nature (IUCN) in consultation with UNEP, the UN's Food and Agricultural Organisation (FAO) and the United Nations Educational, Scientific and Cultural Organisation (UNESCO). The document was circulated widely for comment and improvement. The WCS was eventually published jointly by IUCN, UNEP and WWF with the following objectives:

1   *to maintain essential ecological processes and life support systems such as soil, forests, agriculture and coastal and fresh water systems;*
2   *to preserve genetic diversity in agriculture, forestry and fisheries;*
3   *to ensure the sustainable use of species and ecosystems.*

The priority areas for action were identified as:

1   *law and international assistance – 400 multilateral conventions already existed but few had conservation as their primary purpose;*
2   *more effective management of tropical forests and dry lands;*
3   *a global programme for protecting genetic resource areas;*
4   *more effective management of global commons such as the oceans and atmosphere;*
5   *regional strategies on international rivers and seas.*

During the 1980s many environmental books were produced including *State of the World* 1984 published by the Worldwatch Institute (WI). *State of the World* is an annual publication and is translated into 27 languages. In 1991 *Vital Signs* was introduced and is also published annually giving details of the trends that are shaping our future.[18]

In 1984, *die Grünen*, the West German Green Party, won 7 seats in the European Parliament, having won 27 seats in the West German Bundestag (Parliament) the previous year. In the same year Bhopal, India became famous for the accidental escape of poisonous gas from the Union Carbide factory which led to the death of 2600 people.

In 1986 the Chernobyl nuclear reactor went out of control resulting in clouds of radiation spreading across Northern Europe and Great Britain. Two hundred and fifty people were killed, 100,000 evacuated and thousands of square miles contaminated. Despite a commitment in 1991 to close the three remaining Chernobyl reactors they were still in use in 1996 and much of the surrounding area remains uninhabitable.

In 1987 The Single European Act (SEA) set out the ways in which the European Union should protect the quality of the natural environment, improve human health and contribute to the prudent use of resources. As a result, legislation in the UK and other European countries has been strengthened on clean air, control of pollution, dangerous substances and water quality.

Our Common Future was published in 1987 as a result of the work of the World Commission on Environment and Development, chaired by Gro Harlem Brundtland, the Prime Minister of Norway.[19] This contains the now famous definition of sustainable development which is given in Chapter 4.

In 1988 *The Green Consumer Guide* became a best seller, followed in 1989 with *The Green Consumer's Supermarket Shopping Guide*.[20] On the same theme *The Global Consumer* was published in 1991.[21]

The International Panel on Climate Change (IPCC) was established in 1988 and its work has steadily developed a wide consensus among scientists from all over the world. They have confirmed their belief that human activity is affecting world climate and predicting that sea levels could rise by about one metre during the next century.

The 1990s

Several important books were published in 1990/91 including *The State of the Environment* by the Organisation for Economic Cooperation and Development (OECD).[22] This is an authoritative publication approved by the Environment Ministers of the EU member countries and issued without restriction. The World Conservation Strategy was updated by publication of *Caring for the Earth* by IUCN, UNEP and WWF.[23]

Preparations were being made for the UN Conference on Environment and Development (UNCED) and Maurice Strong was again asked to take the role of Secretary General. Strong asked Stephan Schmidheiny, a successful Swiss industrialist, to prepare a business report for UNCED. This was published prior to the conference entitled *Changing Course*.[24] In the Preface Schmidheiny says:

There is an inescapable logic in the concept of sustainable development . . .We have outlined a change of course for business that can have far-reaching consequences for most business people.

UNCED, which took place in Rio de Janeiro in June 1992, was convened by the UN General Assembly and attended by over 100 heads of state and a further 78 government delegations. Non-Governmental Organisations (NGOs) organised their own Forum some 30 miles away from the conference. It was attended by 500 groups from all over the world. UNCED approved a Framework Agreement on Climate Change, a Convention on Biodiversity, Agenda 21 (the action plan to achieve sustainable development during the twenty-first century), the Rio Declaration and a statement of principles relating to exploitation of forests.[25]

In 1992 the sequel to *Limits to Growth* was published, entitled *Beyond the Limits*.[26] Its conclusion was that in many ways sustainable global limits have been exceeded but there was a note of optimism about the prospects for recovery and lasting solutions. In the same year Al Gore, now Vice President of the USA, published *Earth in the Balance* drawing on his wide experience as a US Senator who had had the opportunity to sit on numerous committees which examined many aspects of the environmental challenge.[27] Gore believes that:

We each need to assess our own relationship to the natural world and renew, at the deepest level of personal integrity, a connection to it. And that can only happen if we renew what is authentic and true in every aspect of our lives.

Climate disturbance and the rising incidence of windstorms are causing growing concern and resulting in increased loss of life and huge economic losses. For example, 15 million trees were felled in Southern England in a single night in October 1987, 120,000 people were killed by a cyclone in Bangladesh in April 1991, many countries have experienced unprecedented weather conditions and Britain had its worst drought for 150 years in 1995. Hurricane Andrew in Florida in August 1992 cost US$20 billion. The effects of this hurricane and Cyclone Iniki resulted in nine insurance businesses going bankrupt.[28] The scientific position on climate change is summarised by Sir John Houghton, co-chairman of the Science Assessment Working Group of the IPCC in his book *Global Warming: The Complete Briefing*.[29] He states:

Scientists are confident about the fact of global warming and climate change due to human activities. However, substantial uncertainty remains about just how large the warming will be and what will be the patterns of change in different parts of the world.

BUSINESS LEADERS GET INVOLVED

The first World Industry Conference for Environmental Management (WICEM 1) took place in Versailles in 1984 organised jointly by the International Chamber of Commerce (ICC) and UNEP. WICEM 2 took place in Rotterdam in 1991 and during this conference the Business Charter for Sustainable Development (BCSD) was agreed. The Charter has 16 clauses about sound environmental management, the last of which commits signatories to assess their organisation's environmental performance and report on the ways in which the preceding 15 clauses have been implemented and are being monitored.[30]

The ICC received reports from many of the 1000 companies worldwide which had endorsed the Charter in the first year and published these results in *From Ideas to Action*.[31] The Charter has been translated into 23 languages and over 2,000 companies have now endorsed it. However, not all the signatories have reported progress and several have found that honouring the commitment takes time, effort and perseverance.

In order to provide an appropriate forum for chief executives to meet and agree action on global environmental issues, ICC formed the World Industry Council for the Environment (WICE). For about two years this body worked independently from ICC and the Business Council for Sustainable Development which had been formed by Schmidheiny to develop the report for UNCED. Both WICE and the Business Council for Sustainable Development were trying to attract the same people – chief executives of major businesses from around the world. In January 1995 the two organisations merged to form the World Business Council for Sustainable Development (WBCSD).

The WBCSD has a supplement in every issue of *Tomorrow* magazine.[32] They report progress on projects, describe new initiatives, report on activities by Regional Groups working with WBCSD and publish the list of members of WBCSD. An impressive list of companies are members, including AB Volvo, Aracruz Celulose, BOC Group, British Gas, Ciba Geigy, Deloitte Touche Tohmatsu International, Dow Chemical Company, Du Pont, Glaxo Wellcome, ICI, International Development Center of Japan, Mitsubishi Electric Corporation, National Westminster Bank, Norsk Hydro, Procter & Gamble, Shell International, Thai Farmers Bank, Taiwan Cement Corporation, THORN EMI, and Toyota Motor Corporation.

Organisational change and development requires commitment from the top and the companies which have become members of WBCSD are publicly declaring that the chief executive is committed to sustainable development. This commitment alone will not achieve change but without it the resources provided for the relevant work to be done would be restricted or cut. A great many companies have already achieved significant reductions in their waste and reduced emissions to the environment, thereby often saving considerable sums of money.[33] Several companies have set up appropriate environmental management procedures and some are actively working towards a better understanding of sustainable development and its implications for their business. More details are given in later chapters.

The UK government set up the Advisory Committee for Business and the Environment (ACBE) in 1991 to strengthen the dialogue between government and industry. It is jointly sponsored by the Department of Industry and the Department of the Environment. Since its formation it has been reconstituted twice and its remit revised so that it can focus on topical issues. The committee is chaired by a chief executive and all the members occupy similar positions from a wide cross-section of industry. The ACBE has published six reports, each of which contains a response from the Secretary of State for the Environment and the President of the Board of Trade. The most recent report includes a summary of recommendations, the government response and an update on progress.[34] Sir Anthony Cleaver, chairman of UK Atomic Energy Technology and Business in the Environment, former chairman of IBM UK and the RSA Tomorrow's Company Enquiry, comments:[35]

British Industry has made massive progress on environmental issues within a (mostly) voluntary framework. And it is working...[in] the US ... Environmental laws were brought in without sufficient business consultation. The result? An adversarial approach between legislator and entrepreneur.

Business in the Community (BiC) was formed by Prince Charles and later added Business in the Environment (BiE).[36] Both tend to focus on small and medium size businesses and involve many senior executives from different industries.

Following Charles Handy's Shank Memorial Lecture to Fellows of the RSA in 1991, an inquiry was launched entitled Tomorrow's Company: the Role of Business in a Changing World.[37] An Inquiry Team was set up involving some 25 companies represented by their top executives. The inquiry lasted two years and also involved some 400 RSA Fellows who were interested in the subject. (See Chapter 3.)

Market and Opinion Research International (MORI) have investigated the values and attitudes of 'Captains of Industry' and the findings suggest that British companies do not pay enough attention to the environment, but regard environmental responsibility as a most important issue. Top executives pay increasing attention to the environment because, as individuals:[38]

- their green values may be deeper than their concerns about the profitability of their companies;
- they recognise that they will need to respond to stricter legislation; and
- they are concerned about the long term decline in public confidence in business organisations.

A survey at the Strategic Planning Society annual conference in 1995, answered by 40 top executives, indicated that two-thirds of those who replied agreed with the following statements about their own behaviour at home:[39]

- we buy and consume more than we need;
- protecting the environment will require most of us to make major changes in the way we live; and
- we take environmental factors into account when making buying decisions.

Surveys of this kind suggest that although many top executives are well informed about the environmental challenge and the implications of sustainable development they are reluctant to be open about these beliefs at meetings with other top executives.[40] The result is that they have less impact on company policy than they could have. The reluctance to speak up is not surprising because the prevailing organisation culture in most companies leads to unconventional views being challenged. However, in the right circumstances, with one or two allies, the risk of introducing new thinking may be less than is imagined.

Opening up a new subject as large as culture change, based on concern for the natural environment, may seem like a daunting prospect. It is certainly not a matter to be taken lightly, nor is it something that can be achieved quickly by a statement from the chief executive. Organisation cultures can be changed but this requires a systematic approach, with commitment and perseverance. Top management also need to set the example by behaving in ways that are consistent with the desired culture.[41]

The financial sector is often criticised for being particularly opposed to environmental matters, treating the subject as largely irrelevant to the performance of business, but that is changing. The National Westminster Bank (NatWest) in the UK has been a leading exponent of environmental management and Derek Wanless, Chief Executive of NatWest, has been a member of ACBE since it was formed in 1991 and was its chairman until it was reformed in 1996. In the USA Salomon Inc has been described as the first Wall Street firm to green its core business.[42] Robert Denham, CEO says:

Environmental issues are very important to a lot of industries we're involved with and if we can have a better understanding of those issues and deliver better advisory services in transactions or acquisitions, then that's another factor to distinguish us from our competitors.

MAJOR THREATS TO THE EARTH[43]

A brief summary of some of the major threats and the problems which need to be tackled if sustainable development is to be achieved are given below. Those in business who believe that environmental protection only entails costs and restrictions are mistaken. The search for solutions started with controlling pollution where it occurred but quickly moved to preventing it at source – because it made better business sense – and then broadened to include efficient use of resources. This often stimulated the search for new technology, the use of different materials and better design. The need for creativity and imagination was based on a sound understanding of the principles of ecology and sustainable development.

Developing countries need help in avoiding the wasteful, polluting and outdated methods of the industrialised countries and this calls for creative ways to transfer technology and find new solutions. All of this creates business opportunities which will only be realised if the strategic thinking in many different industries recognises what building to last means.

Twelve of the most important environmental problems are:

1 Global warming and climate change–carbon emissions reached an all time high in 1996, with rapid increases in developing countries; seven of the eight warmest years during the past century occurred between 1980 and 1992.[44]

2 The increasing disparity in wealth between industrialised and developing countries. There are strong links between environmental degradation, poverty and population growth.

3 Population growth – world population is growing by 87 million each year – that is over seven million people more to feed every month. Over 20 million refugees are receiving UN assistance and the numbers are growing each year. Urbanisation continues to rise rapidly.

4 World grain production remains steady and world grain stocks dropped to their lowest level ever in 1996 with 48 days supply available. The land area used for harvesting grain worldwide has been dropping since 1989. Soil erosion and intensive farming practices aggravate the problems.

5 Overall grain consumption per person, when taking account of grain eaten by people and by animals which are then eaten, varies greatly worldwide. The average in the USA is 800kg per person per year, in Italy 400kg, in China 300kg and in India 200kg. The higher levels of grain consumption represent diets with more meat. The trend in countries like China is to eat more meat and China is ceasing to be self sufficient in grain production, which could have worldwide repercussions on grain prices.[45]

6 Water shortages and the build up of salt in 10 per cent of the world's irrigated areas results in lower crop yields. In places where water shortage occurs frequently there are life threatening consequences.

7 Ozone depletion – while CFC production continues to drop the trade in illicit CFCs appears to thrive.

8 Reduction in biological diversity which is leading to simplified, more vulnerable ecosystems.

9 Resource depletion – renewable resources such as timber and fish are being depleted at unsustainable rates; non-renewable resources such as fossil fuels, and some metals and minerals are in urgent need of conserving.

10 Transport, especially road freight, use of cars and air travel deplete resources and add to air pollution.

11 About half the tropical forests of the world have been felled in the last 50 years and the destruction continues. The land is required to grow food for local consumption and for sale to raise foreign exchange. The implications affect all humanity.

12 Resource shortages and disputes over environmental issues add to social and political tension in many parts of the world.

The manifestation of severe problems is obvious in a severely degraded natural environment but this is not the only evidence that is important. Sometimes the underlying cause of social or economic turmoil could be that the community is pressing against the limits of a sustainable way of life. For example Lester Brown of the Worldwatch Institute has said:

The history of the next few decades will be defined by food, specifically by rising prices of both oceanic and land-based food products, by a spreading politics of food scarcity, and by an increasingly intense struggle to achieve a sustainable balance between food and people.[46]

It is important to realise that the carrying capacity of the environment can be exceeded for several years by people living off the 'capital' stock of the natural world but eventually the collapse will occur. The analogy with business practices is helpful. Companies which erode their capital know this cannot continue for long. There is no equivalent accounting process which reveals the erosion of the Earth's natural capital, but ways to show this are being explored.[47]

A COMMON PREDICAMENT, PURPOSE AND PLAN

No single country can solve the global crisis, nor insulate itself from the consequences – hence the vital role of the UN and the frequent attempts to achieve international agreements through conventions, protocols and treaties. In *Earth in the Balance*, Al Gore, before becoming Vice President of the USA, proposed a Global Plan, based on the successful Marshall Plan at the end of World War II, which has recently been endorsed by Sir John Houghton of IPCC in his book *Global Warming* (see note 29). The plan includes six strategic goals:

1 the stabilisation of world population;
2 the rapid creation and development of environmentally appropriate technologies;
3 a comprehensive and ubiquitous change in the economic 'rules of the road' by which we measure the impact of our decisions on the environment;
4 the negotiation and approval of a new generation of international agreements that are sensitive to the vast differences of capability and need between developed and developing nations;
5 the establishment of a cooperative plan for educating the world's citizens about our global environment; and
6 the establishment of social and political conditions most conducive to the emergence of sustainable societies.

The UN has set up the Sustainable Development Commission, and the governments in the USA and the UK, among others, have established their own equivalent bodies. Round tables have been formed in Canada, USA, UK, France, The Netherlands, China and India to provide a broadly based forum to examine the issues and debate solutions. The UK Government Panel on Sustainable Development has been established. The responsibility to take action has been delegated to another organisation called 'Going for Green'. Other countries have their counterparts to organisations like these.

Al Gore stresses the importance of commitment by large numbers of people and the importance of shared goals:

Whether we realise it or not we are engaged in an epic battle to right the balance of our Earth, and the tide of this battle will turn only when the majority of people in the world become sufficiently aroused by a shared sense of urgent danger to join an all-out effort ...Though it has never yet been accomplished on a global scale, the establishment of a single shared goal as the central organising principle for every institution in society has been realised by free nations several times in modern history.[48]

Effective businesses, whatever their size, have demonstrated time and again how to achieve an alignment of common effort. The skills which accomplish this in business have never been more urgently needed, but now in a wider context. The role of business in achieving sustainable development is vital.

CONCLUSION

This chapter has provided a brief overview of the evolution of environmental concern worldwide. It has shown that ecological and social problems are inter-related and can only be solved within the framework of a healthy economy and a peaceful world. The need for new thinking is demonstrated by the concept of carrying capacity and the growing realisation that conventional thinking needs to evolve into a science-based radical philosophy, with a strong ethical foundation, to guide the development of national and international policies in the coming years.

Business leaders are already involved in the search for solutions and in helping to shape the guidelines for practical action. However, those who recognise the dangers of ecological ignorance and social denial need to prevail over others who continue to argue for slowing down the transition to sustainability.

A brief summary of twelve of the most serious problems is provided in order to highlight the priorities which need attention.

Finally Al Gore, Vice President of the USA, suggests some practical ways to move forward in the search for lasting solutions. He articulates the basis for a global common purpose so that international institutions, national governments, businesses and individuals can all play their part.

2 CHANGE
WITHIN COMMUNITIES

During the last 50 years the main political goal in most countries has been to raise material standards of living. This emphasises the need to maintain high levels of employment. These aims have been reinforced with people seeking to improve their lives and politicians promising to achieve what the people want. Government promises have only partially been achieved.

In addition to the pursuit of material ambitions people have also sought more social justice, improved education and better health care. Many special interest groups have been formed to pursue special needs. When their aims remain unfulfilled people seek more involvement in decisions that affect their lives. Increasingly people are forming coalitions or alliances to work together to strengthen their influence.

Alongside these social aims the pace of life continues to accelerate. Worldwide travel, technological advance, and improved communications create a complex and confusing background for social goals. The world appears to become smaller and more fragmented. Social cohesion breaks down and for many disruption of family life occurs. Fifty years ago the social stresses were different with lower standards of living, relatively stable family life but more conflict between nations.

When the acquisition of new knowledge was relatively slow people could adhere to a set of beliefs throughout their life. This gave rise to some conflict between generations but generally most people coped with the changing values of grandparents, parents and children. With the accumulation of knowledge people found that they needed to change some of their values and ideas in order to keep pace with the times. For many people this may now happen once or twice within their life span. Families can be disrupted when the two partners are not learning the same things or at the same speed. They may wish to change their values and lifestyles at different times and in different ways. The same things can occur at work where conflict may arise but the issues are either resolved or people leave to work elsewhere.

At work there have been other changes. Most large organisations which used to provide lifetime careers no longer do so. In recent years there have been massive reductions in the number of people employed in many of the largest companies. This is stimulated by improved technology and cost cutting in order to maintain profits. Some now talk of anorexic companies which have cut back too far. Many of those who leave big organisations work alone or form small companies which are now the main area for new employment opportunities.

Against this general background it is not surprising that many people have become disillusioned and sought their own ways to improve their situation. This takes many forms and as yet only affects a small minority, but the diversity of initiatives is impressive. Some of the new thinking and original solutions which people have worked out for themselves are described below. They would need to develop rapidly before they become significant and this may never happen, but they may contain ideas that are relevant for future sustainable societies. Without these there will be no lasting success for business.

Gross National Product (GNP) is the accepted, traditional way in which wealth and national progress is measured. It is generally assumed that countries with a rising GNP are meeting the needs of their people. However, GNP excludes many social and environmental factors which are recognised as important so it is not surprising that the over reliance on GNP as the universal measure of progress is now being questioned.

GNP AND THE INDEX OF SUSTAINABLE ECONOMIC WELFARE (ISEW)

GNP measures wealth and economic progress. It is not a good indicator of national well-being because it does not take account of social factors, environmental damage and depletion of resources. A successful economy has increased trade, which usually means more motorised transport, more roads and higher emissions of pollution especially in urban areas. However, when this results in one in seven children suffering from asthma (for example, in London, UK) the medical treatment these children receive increases GNP. If there are more road accidents resulting from increased traffic, all the services required to deal with the accidents are added to GNP. Development, which is widely associated with progress often means clearing away the natural environment, building or widening roads and increasing urbanisation. Many people are fearful when they learn about 'development' close to their own homes.

GNP rises with increased trade. This means that as 'do-it-yourself' (DIY) grows in popularity the materials used are included in GNP figures but labour is not. Similarly if people stop growing their own vegetables and cleaning their own homes but pay others to do this for them, GNP rises. Likewise those who start growing their own vegetables, repairing or decorating their home and washing their own car are reducing GNP. When solar panels are fitted to heat water only the purchase of the panels is added to GNP; the energy savings are deducted. These examples highlight the weakness in relying so strongly on GNP as an indicator of social welfare and progress.

The New Economics Foundation (NEF) has undertaken a detailed study of the Index of Sustainable Economic Welfare (ISEW) and applied their findings to the UK for the period 1950 to 1990. The ISEW aims to assess the quality of life over time. This is done by deducting expenditure which is incurred in repairing social and environmental damage from GNP. The long-term cost of environmental damage and depreciation of natural capital is taken into account. Changes in the distribution of income are included because £1 means more to the poor than to the rich. Housework, which is excluded from GNP, is included to allow for the non-monetary benefits to people and the economy. A full table of the adjustments to GNP are available from NEF.[1]

The result of these calculations produces the ISEW for the UK and it is compared with GNP over the period 1950 to 1990 in the graph shown in Figure 2.1. From the 1970s onwards ISEW has been declining even though GNP has continued to rise. Since 1950, long-term environmental damage has increased steadily, as has the value of household labour. Factors like these have produced the drop in ISEW.

This analysis provides powerful evidence of the need for change but does not show that it is already happening. It seems plausible that the 'feel good factor' remains illusive because ISEW is declining.

Figure 2.1 GNP and Index of Sustainable Economic Welfare (ISEW) per capita in 1985
£ sterling

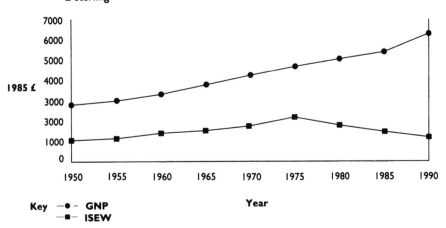

LOW PROFILE TRENDS

Many people are dissatisfied with society as it is and a growing minority in countries
like the UK are unwilling to accept traditional lifestyles without question. Some have
started to experiment with new ways of living or changing important aspects of their
lives. In several areas these are becoming sufficiently significant to be measurable.
They indicate the emergence of new trends – as yet not well known. It is interesting
to examine the following:

- traditional beliefs contrasted with emerging ideas;
- ethical and environmental investments;
- energy efficiency and renewable energy;
- householder initiatives;
- diet and health;
- organic farming and food;
- alliances and The Real World Coalition;
- communications; and
- Charter 88 and the campaign for a new UK constitution.

TRADITIONAL BELIEFS CONTRASTED WITH EMERGING IDEAS

Since the Industrial Revolution conventional values have been widely accepted and
helped the evolution of society, especially in Europe and North America. These values
are now being challenged by new beliefs broadly emanating from the strengthening of
the environmental movement over the past 20 to 30 years. A brief summary of these
contrasting values is given in Table 2.1.[2]

Table 2.1 Traditional Beliefs and Emerging Ideas: Industrial Countries

Traditional beliefs	Emerging ideas
• By felling forests we get more land for agriculture.	• Denuded forests deplete natural resources causing more problems than they solve.
• By going further out to sea we can increase fish catches.	• If we do not conserve fishing grounds there will be no fish.
• Chemical fertilisers and pest control increase food supplies.	• Organically grown food is better for people's health and ecosystems.
• Species have always disappeared through natural processes, there is no need to worry.	• Rapid reduction in biodiversity may threaten the planet's stability as well as eliminating potentially useful species.
• Plentiful energy, especially fossil fuels, are essential.	• Energy conservation measures and renewable energy sources help the environment and provide new business opportunities.
• Science and technology provide improved standards of living and indicate progress.	• Human scale and appropriate technology which is democratically controlled is the better way to improve quality of life.
• Population growth is a natural phenomenon, often desirable.	• Arresting population growth is essential. Improved health care, education, family planning services and relief of poverty all help achieve this goal.
• Economic growth benefits all and can go on for ever. GNP is an effective measure of progress.	• New indicators are showing that sustainable human welfare is not the same as economic growth measured by GNP.
• As the rich get richer the poor will also benefit.	• Poverty relief requires direct action – helping people to help themselves.
• The arms trade is a legitimate export business, maintains jobs and helps the economy.	• A more peaceful world needs fewer, perhaps no weapons; the money and resources can be much better used elsewhere.
• Economic criteria are the best guide to decision making in business.	• Economic criteria need to be complemented by social and ecological measures for sound business decisions.

The new millennium is signalling a need for new thinking which encourages emerging ideas such as those in Table 2.1. New values will form the basis of a desirable future society which differs from existing societies in significant ways.

New measurements will be needed to assess progress towards a better quality of life and towards sustainable societies. The interest in ISEW is an example of a new measure which could be very useful.

ETHICAL AND ENVIRONMENTAL INVESTMENTS

Ethical and environmental investments have shown steady growth in the USA and the UK. Ethical investments tend to avoid specific companies which are deemed to have unethical products or services, while environmental investments tend to select those companies which are taking positive action to help the environment. Holden Meehan have published reports on progress since October 1991 and their most recent guide was published in 1996.[3]

The growth in these investments in the UK has been impressive. In 1990, £280 million was invested. This had risen to over £1000 million by 1996. Figure 2.2 shows the growth from January 1989 to July 1996.

Figure 2.2 Ethical Investments in the UK[4]

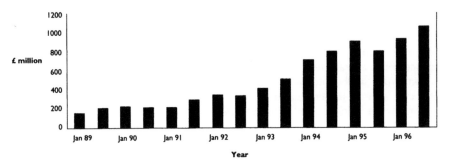

There are 35 Ethical Unit Trust schemes in the UK and a Green Independent Financial Advisers organisation has been formed. The Ethical Investment Research Service (EIRIS) provides a detailed guide to these investments which cover pensions, Personal Equity Plans (PEPs), endowment mortgages and other ethical investment plans. Their publication *Money & Ethics*[5] describes 28 of the leading ethical funds, the product range covered by each and a comprehensive analysis of what each fund avoids as well as the types of business activity each supports. Holden Meehan in their guide assess the performance of each fund in terms of:

- ethical criteria;
- environmental criteria;
- the pedigree of the fund manager;
- the launch date and total sum invested to date;
- geographical spread;
- resources applied to screening; and
- products available from each fund.

Although some investors are satisfied with a reasonable return others wish to know that these investments compare favourably with those which are screened only in terms of their financial performance. Holden Meehan believe that there should be 'no special pleading' for ethical investments and the figures in Table 2.2 show comparisons for 1995.

Table 2.2 Comparative Performance of Unit Trusts 1994/95

Description	Growth %
UK All-Share Index growth	20.14%
FTSE 100 Share Index growth	21.75%
Hoare Govett Smaller Companies Index growth	12.3%
Sector average growth rate for comparable funds	14.2%
Ethical funds growth rate	14.94%

The general prediction in 1996 is that smaller companies will perform better than larger companies and this will encourage more ethical investment. This seems to contradict the general finding that smaller companies do less to reduce their environmental impact than larger companies. However, when it comes to ethical investment it is possible to select those companies whose business is financially sound and ethically or environmentally responsible in a growing sector of the market. These companies should perform well. Holden Meehan conclude that 'ethical funds can produce above average returns over meaningful time periods.'

Unit trusts on the international scene had an average growth rate of 6.5 per cent compared with 11.48 per cent for ethical unit trusts. New opportunities for ethical investments are being created. For example the Environmental Value Fund has been launched in Norway. It includes the most environmentally responsible companies in nearly every sector, rather than excluding whole industries because they cause environmental damage.[6] Kleinwort Benson Investment Management have launched a fund which selects 'Tomorrow's Companies' using the criteria developed in the RSA's Tomorrow's Company Report (see Chapter 3).[7]

ENERGY EFFICIENCY AND RENEWABLE ENERGY[8]

Significant changes are taking place in energy trends worldwide as stricter regulations require reductions in emissions of carbon dioxide. The strong public opposition to nuclear power is making it extremely difficult to secure sights for new reactors and old reactors are being shut down. Meanwhile the price of renewable energy continues to drop as new technology makes it increasingly feasible to develop energy derived from sun, wind and tides.

Conventional Fuels

The consumption of coal, oil and natural gas in industrialised countries is diminishing but continues to increase in countries like China where rapid economic expansion is taking place. The overall increase in consumption worldwide of these fossil fuels has been around 1 per cent per annum over recent years. This is due to energy efficiency measures which reduce consumption and the polluting effects of burning these fuels. In the years to come the increased use of renewable energy will deplete further the demand for traditional fuels.

The net installed electrical generating capacity of nuclear power plants has grown very slowly from 310,000 megawats in 1988 to 340,000 megawats in 1995. Following the Three Mile Island accident in the USA (1979) and the Chernobyl accident in the Ukraine (1986) nuclear power has proved extremely difficult to develop. This is due to the cost of more stringent safety regulations and the lack of public confidence in this form of electrical power generation. Global figures take account of new reactors in India, Japan, South Korea, Ukraine and the UK being offset by the closure of old reactors. By the end of 1995 only three reactors were under construction in the Americas – one each in Argentina, Brazil and the USA. Nuclear expansion has stopped throughout Europe with the exception of France where four reactors are under construction. China signed a deal in 1995 for two reactors to be supplied by France, with heavy French government subsidies.

Renewable Energy

The use of renewable energy (wind and solar especially) is growing rapidly. This is the result of new technology, the drop in prices of the equipment that is required and switching from the combustion of fossil fuels to harnessing the natural forces of the sun and wind. The figures for recent years are given in Table 2.3.

These are phenomenal growth rates. Any business would welcome a sales growth of 12 per cent annually – doubling in six years. The figures show a growth rate even faster than that.

The desire to save energy has led to the development of compact fluorescent light bulbs. Demand for them has been rising rapidly. Although these bulbs cost more they last much longer and use a lot less electricity. Using a long life, low energy light bulb in a hall or stairway can save up to £33 during its lifetime if it is in use for four hours a day and electricity costs 7.5p per unit. Replacing all light bulbs which are in use for three or four hours a day can save 20 per cent of electricity costs, cut carbon dioxide emissions and the capital expenditure is recovered within two or three years.[9]

Table 2.3 Renewable Energy Expansion

Year	World wind energy generating capacity (megawatts)	World photovoltaic shipments (megawatts)	World sales of compact fluorescent bulbs (millions)
1981	25	7.8	
1982	90	9.1	
1983	210	17.2	
1984	600	21.5	
1985	1020	22.8	
1986	1270	26.0	
1987	1450	29.2	
1988	1580	33.8	45
1989	1730	40.2	59
1990	1930	46.5	80
1991	2170	55.4	115
1992	2510	57.9	139
1993	2990	60.1	180
1994	3680	69.4	210
1995	4880	81.4	240

The use of recycled materials such as scrap steel and post-consumer waste paper is rising fast. The best source of these materials is urban areas and manufacturing units are now being developed by smaller businesses located close to the source of their raw materials and their markets. For example in the USA scrap steel is used by 'minimills' which now supply 35 per cent of US steel, compared with only 5 per cent in the 1960s. Minimills are small and efficient. By using scrap steel they require less iron ore and less limestone and consume only one-third as much energy as conventional production methods. Minimills are ideal for developing countries with small dispersed markets. Recycled paper accounts for over 80 per cent of the market in Taiwan, Denmark, Mexico and Thailand. Manufacturers are earning higher profits by using post-consumer recycled paper and they are signing contracts with cities to secure their supplies of scrap paper.[10]

HOUSEHOLDER INITIATIVES

In many UK communities the strongest shopping trend appears to be the desire to use out-of-town supermarkets and ignore the decline of local producers and smaller shops in towns and villages. The extra traffic this generates and the demise of small traders has led to protests and the government has stated its intention to discourage any further development of large out-of-town supermarkets. In addition to government action there are several ways in which people are finding their own solutions to seemingly intractable problems. Householder initiatives include:

- Cooperatives, housing associations and credit unions;
- Local Exchange Trading Systems (LETS); and
- Global Action Plan for the Earth (GAP).

Cooperatives, Housing Associations and Credit Unions

This is often referred to as the social economy where the primary purpose is to meet a need rather than provide profit to shareholders. Cooperatives and housing associations are organisations which are professionally run by volunteer boards of management. Community enterprise of this kind has been growing steadily since the mid 1970s and has proved invaluable in poorer and more remote areas. Many of the people who benefit were unable to raise even small loans or get support for their initiatives until the advent of community enterprise.

Credit unions make money available to people in local communities at low interest rates to enable them to undertake initiatives which require some capital investment but which high street banks are unwilling to consider. In essence a credit union is a 'save and borrow' cooperative. The members save on a regular basis, then borrow from the common fund when the need arises. The cynical belief that poor people would not save and would fail to repay their debts has not been borne out.

There are now 500 credit unions in the UK and 30 are in Birmingham where the Birmingham Credit Union Development Agency has been particularly successful.[11] The New Economics Foundation has published a paper on community banking which provides a summary of worldwide initiatives.[12] These schemes help people, often in deprived areas, by making money available at low interest rates. The only alternative for many people would be to borrow from loan sharks, who charge exorbitant rates of interest, because conventional banks regard people in deprived areas with few if any assets as unacceptable risks.

Local Exchange Trading Systems (LETS)

LETS were started in Vancouver, Canada by Michael Linton. He was searching for some way to help the unemployed, which included himself.[13] After trying without success to barter his own skills, Linton decided that a new kind of money was needed and this gave birth to LETS. He invented a 'currency' which facilitated local trade but could not be invested, which was community-wide and used for barter trade on a non-profit basis. Originally the currency was called the 'Green Dollar' but as the LETS idea has spread local communities have invented their own names so in the UK there are now 'Strouds' in Stroud, 'Brights' in Brighton, and 'Olivers' in Bath.

The essence of the scheme is that people who want to trade together make their own arrangements with the help of a starter pack, which in the UK is provided by the UK Letslink office.[14] In each locality where a scheme is initiated someone needs to keep a list of participants and an account of transactions in the local LETS currency. Typically each unit of currency is the equivalent of £1 and people wishing to trade determine the price for their goods and services at their discretion. If they get swamped by demand their price could be too low, or no demand might suggest an unwanted product or service or that the price is too high. The guiding principles for LETS are:

- someone needs to represent the 'agency' and keep a system of accounts;
- people who join the scheme pay a small annual membership fee;
- no money is deposited or issued, all accounts start at zero;
- the agency acts on authority from the account holder in making credit transactions from one account to another;
- there is never any obligation to trade;
- account holders may know the balance and turnover of others;
- no interest is charged or paid on balances;
- administration costs are recovered, using the local currency, from accounts on a cost service basis;
- a catalogue of members of each scheme is drawn up as a trading sheet, showing 'wants' and 'offers', contact names and telephone numbers;
- individuals and businesses can participate; and
- 'payments' for goods and services may be partly in the local LETS currency and partly in sterling.

LETS benefits include the following:

- creates some employment;
- enables unemployed people to recognise that their skills and experience are of value to others and thereby regain some self respect;
- helps build community spirit;
- improves the quality of life especially for people with low incomes;
- liberates goods and services within the local community;
- creates a local currency which complements money transactions; and
- reduces the flow of money leaving a community or locality.

It is not easy to find out how successful LETS have been worldwide. There is some indication that they have been particularly successful in the UK and Australia but have not spread rapidly in Canada and the USA. In the UK the Letslink office is swamped by enquiries to start new schemes. The growth of schemes has been striking as is shown by the figures in Table 2.4.

Table 2.4: Local Exchange Trading Schemes in the UK

Year	1990	1991	1992	1993	1994	1995
Number of schemes	5	10	15	35	200	400

It is estimated that in 1995 there were 30,000 people in the UK taking part in LETS. This figure is not easy to estimate because after a while two people who provide a service to each other on a regular basis may not record every transaction through LETS. For example one family may want a baby sitting service in exchange for working in a neighbour's garden. There may also be some turnover of participation in LETS as individual circumstances change. Some schemes struggle to get started and a few fail. Despite this there is little doubt that LETS show a progressive trend and some local councils are appointing LETS officers to develop schemes. These include Cardiff, Greenwich, Hounslow, Liverpool and Manchester in the UK.

The question of taxation on LETS transactions has been raised and the LETS response is that this is a reasonable question. Their answer is that if it is really necessary then it should be paid in the local currency. For example, by cleaning the windows of the council offices or providing some other service to a government department! Any attempt to make tax a formal requirement could result in more transactions reverting to barter deals which are not recorded. Furthermore, the benefits of LETS, as listed above, suggest that these schemes have desirable characteristics for sustainable communities.[15]

Global Action Plan for the Earth (GAP)

GAP started in the USA after Earth Day 1990. GAP is an international charity now being developed in 10 countries in North America and Europe with exploratory discussions taking place in Japan and Korea. GAP's mission statement is:[16]

to encourage individuals to take effective environmental action in their homes, workplace and communities.

Table 2.5 Global Action Plan – Performance Figures 1996

Schemes were started at different times. The cumulative figures are shown below. Savings are reported by households based on their own measurements, then aggregated. Verification is by sampling.

Country	Number of Households	Garbage savings %	Water savings %	Home energy savings %	Transport fuel savings %
Belgium	610	N/A	N/A	21	N/A
Denmark	186	49	18	37	N/A
Finland	218	54	22	9	23
Ireland	65	N/A	N/A	N/A	N/A
Netherlands	4907	24	12	12	2
Poland	632	N/A	N/A	N/A	N/A
Sweden	1621	42	18	81	3
Switzerland	741	19	14	71	2
UK	10,043	28	19	9	4
USA	3471	35	41	15	27

Note: N/A means data not available.

The household programme offers six action packs covering waste, energy, water, transport, shopping and next steps. Each month participants note what they are doing currently, read the ideas in the Action Pack, take action as appropriate, then let the GAP office have details of their savings. A summary of collective achievements is sent to all participants so they know how, with others, they are contributing to an improved environment. News about GAP and the national results are published in a quarterly newsletter.[17]

GAP UK began in 1992 with support from the Department of the Environment, companies, local authorities, voluntary organisations, charities and individuals. GAP UK has recruited more households than the schemes in other countries as the figures in Table 2.5 show

In the UK individual households join GAP but in other countries the emphasis is on forming 'EcoTeams' so that a group of households work together, using an 'EcoTeam Workbook' rather than Action Packs. In the UK individual households join the scheme because a strong resistance to 'EcoTeams' was encountered. The best ways to get people involved for lasting results and to assess results is the subject of continuing research. Results to date suggest the six month scheme is sufficient to change habits which tend to continue after the scheme has finished. A significant proportion of those who complete the scheme want to remain involved and encourage others to join.

The household scheme in the UK is complemented by other initiatives:

- working with the Worldwide Fund for Nature (WWF) through their established groups, using the title 'Action at Home';
- working with several local authorities to help them achieve that part of their waste reduction and recycling targets attributable to households;
- establishing 'Action at Work' based on the same principles as the household programme; and
- starting work on a schools programme to involve children in positive action to improve their environment at school and at home.

The GAP UK programme was tested out by The Consumers' Association who got five very different households in Devon, London (2), Scotland and Gloucestershire, to take part in the full programme but spread over only two months. The findings are reported in *Which?* Magazine.[18] They found that motivated people given detailed information about how to change their lifestyles are able to make some changes while other changes prove difficult if not impossible. It is hard to reduce dependency on the car where public transport is inadequate and cycling on narrow roads is dangerous, but one couple achieved considerable savings with planned journeys and car sharing.

After doing this survey *Which?* proposed a helpful definition of sustainable development for consumers:

To achieve a level of consumption that meets the present and future needs and aspirations of people the world over, from all sectors of society, without compromising the sustainability of the environment and its life support systems.

DIET AND HEALTH

Two ways in which people are actively seeking new lifestyles is by changing their diet and giving more attention to their health. Although typical behaviour in developed countries indicates that the majority buy convenience foods, abandon regular meals in favour of 'grazing' on snacks, and do little to safeguard their health, there is a small but growing minority trying out new approaches.

Vegetarianism

A vegetarian diet, or perhaps more accurately for many, a meat-free diet, is now preferred by over five million people in the UK and this figure is increasing. Red meat consumption was being avoided by growing numbers of people before the beef crisis in spring 1996. Since the BSE scare one million people in the UK have turned to a vegetarian diet, and the Vegetarian Society has been sending out 200 information packs a day.[19] Several organisations have monitored the growth of vegetarianism in its various forms since the early 1980s and a summary of the research is given in Table 2.6.[20]

Table 2.6 Vegetarianism

% of Population	1984	1986	1988	1990	1993	1995
Vegetarians	2.1	2.7	3.0	3.7	4.3	4.5
Avoid red meat	1.9	3.1	5.5	6.3	6.5	7.3

Mintel surveys have been carried out periodically during recent years and the findings include the following:

- 11 per cent of 15–24 year olds do not eat red meat (July 1990);
- 8 per cent of adults avoid red meat or are vegetarians (July 1991);
- 10 per cent of adults have given up or avoid red meat (September 1991);
- 10 per cent of 15–19 year olds are vegetarians (October 1991);
- 10 per cent of single adults under 54 have a mainly vegetarian diet (September 1992);
- retail sales of 'vegetarian food' were worth £11.2 billion in 1992;
- sales of vegetarian-specific food were valued at £349 million in 1994 and expected to grow to £398 million in 1995 and £603 million by 1999 (October 1995);
- the 'vegetarian' market grew by 79 per cent from 1990 to 1994; and
- since 1990 30–40 per cent of respondents have claimed to eat less red meat.

The reasons for eating more vegetarian meals or becoming a vegetarian vary. Some people dislike the idea of killing and eating animals, others believe the diet suits them and some recognise that it has environmental benefits. More people can be fed if grain is eaten by humans rather than fed to animals which are then eaten.

Producing meat needs varying amounts of grain. It takes 7 kg of grain to produce 1 kg of beef, 4 kg of grain to produce 1 kg of pork and 2 kg of grain to produce 1 kg of chicken. Recent calculations suggest that if everyone in the world ate the same diet as Americans it would be possible to feed only half the current world population. A diet comparable to the Italian diet could feed the current world population. The Indian diet, with high consumption of grain, would enable twice the current world population to be fed.[21]

Complementary Medicine

An increasing number of people are exploring the meaning of a holistic approach which places more emphasis on staying healthy rather than curing illness. This means that individuals take more responsibility for their lifestyle, type of work they do, diet, exercise, avoidance of excessive stress and more balance in the their lifestyles. Those taking care of themselves are in a striking contrast to others who pursue stressful lifestyles, rely on convenience foods and take little exercise. The former are more likely to be compatible with a sustainable society.

In recent years complementary medicine has become more acceptable to the medical profession. Complementary medicine places less reliance on drugs to cure symptoms by making more use of natural remedies, therapeutic treatments and prevention of illness. Many people who have tried a complementary medicine in some form regard it as helpful. The use of complementary medicine is not only increasing but is well regarded by most of those who try it.

Figure 2.3 Patients' Verdicts on Complementary Medicine

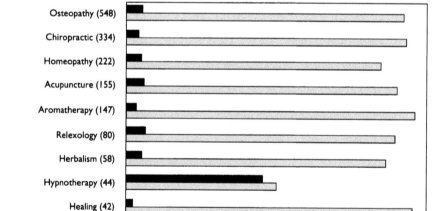

A survey was carried among users of different kinds of complementary medicine. In Figure 2.3 the numbers in brackets indicate the quantity of people in the survey who had used each of the therapies listed. The columns on the bar chart indicate the degree of satisfaction and dissatisfaction for each treatment by those who have tried it. Respondents were also asked to indicate the degree of improvement they experienced, but this is not shown on the chart. Three-quarters of those who have tried healing (some call it spiritual healing) were greatly improved. Half or more were greatly improved by osteopathy, chiropractic and acupuncture, one in three by homeopathy and one in four by aromatherapy and reflexology.[22]

Confirmation that more people are interested in health foods and complementary medicines is supported by the increasing range of products on offer in health food shops and chemists. Many of these products appear to be doing well and this is borne out by the fact that one in three *Which?* members has tried complementary treatments and therapies at some time. In 1994 the total expenditure on over-the-counter 'alternative' medicine was more than £60 million in the UK.

ORGANIC FARMING AND FOOD

Farming is moving slowly towards more sustainable methods including use of less chemical fertilisers, herbicides and pesticides. For 50 years The Soil Association has campaigned for organic farming and has recently merged with the Organic Farmers and Growers.[23] The Soil Association, through its Woodmark scheme, also has an active campaign to encourage responsible forestry which is closely aligned to the Forest Stewardship Council scheme (see Chapter 6).

Organic farming methods are being promoted in many other countries. The sale of organic farm products in the USA trebled in the period 1989–94.[24] In Europe agricultural land used for organically produced food accounts for 7 per cent in Austria, 3 per cent in Sweden, just under 2 per cent in France and Germany but is growing steadily. In Austria, Finland, Greece and Italy the organic acreage doubled in the five year period 1990–95.[25] However, in the UK only 0.3 per cent of farm land is used for organic production but it, too, is slowly increasing. Many inquiries were stimulated by the BSE scare and direct marketing methods are helping to make prices of organically produced goods more competitive.

The Henry Doubleday Research Association is Europe's leading organic research and consultancy organisation with over 17,000 individual members and 200 local authority members.[26] The Sustainable Agriculture, Food and Environment Alliance (SAFE Alliance) campaigns for sustainable agriculture, which is beneficial to the environment and sensitive to consumer demand.[27]

ALLIANCES AND THE REAL WORLD COALITION

An interesting trend in society is the degree to which different organisations are cooperating with each other in order to achieve goals which reflect their common interest. One example of this is the formation of The Real World Coalition and publication of *The Politics of the Real World*.[28] The Real World Coalition comprises

30 organisations with different ambitions. Their specialised knowledge is used to good effect to make a series of powerful statements. Their collective membership totals more than 2,000,000 people. The diversity and strength of this coalition can be appreciated from their membership many of whom are listed in Table 2.7.

Table 2.7 Some Members of the Real World Coalition

Town and Country Planning Association	The Poverty Alliance	Christian Aid
World Wide Fund for Nature (WWF)	Population Concern	Charter 88
New Economics Foundation	Forum for the Future	Oxfam
Save the Children Fund	Friends of the Earth	Transport2000
Black Environment Network	Sustrans	Alarm UK
Employment Policy Institute	UN Association	CAFOD
Medical Action for Global Security	Media Natura Trust	KAIROS

The Coalition's well-reasoned statement covers topics such as global warming, population trends, sustainable consumption, transport, farming and food, poverty, the effects of GATT, the arms trade, income and inequality, unemployment, redefining wealth, and tax reform.

COMMUNICATIONS

The use of e-mail and the Internet has risen meteorically since 1995. In fact it is so rapid that no one really knows how quickly it is growing, how many people who sign up make effective use of it and what the implications are likely to be in the next few years. No one would suggest that telecommunications can replace face-to-face meetings, but it is reasonable to believe that at least some meetings could be replaced by telephone calls, teleconferencing and the use of electronic mail. If this happened it could easily reduce the need for some journeys which would save fuel, cut emissions of carbon dioxide (CO_2) and reduce congestion. The main effect of computers and electronic communications has been to speed up the exchange of messages rather than to create the paperless office. That requires considerable confidence in electronic equipment.

The growth of the Internet seems to be confirmed from various sources such as the number of servers, sales of equipment, the growth of homes with computers and the interest from larger organisations. It has been estimated that the market will soon be worth £200 million a year, that America Online (AOL) is reputed to be attracting 1,000,000 users every 80 days and CompuServe 10,000 a week.[29] How many of these persevere and stay on-line is not known.

British Telecommunications (BT) has pointed out the potential energy savings of using the telephone rather than attending meetings. It has calculated that if a person travels alone by car to a meeting 10 kilometres away, that person could have 21 hours of continuous telephone conversation for the equivalent amount of energy. Similarly if a person travelled, with others, by rail or air to more distant meetings their share of energy consumption would be the equivalent of extremely long telephone conversations. Obviously from BT's point of view this is good marketing but if a few more telephone conversations replaced journeys the environment would benefit.[30]

As the market evolves and technology improves it will be interesting to see whether the big organisations or their smaller competitors gain most. The former have size, customer lists and capital on their side but smaller organisations are more agile, creative and responsive to changing customer requirements. They are probably better equipped to remain focused and to benefit from new technology and new ideas. Big organisations are better able to make investments in both people and technology for larger scale projects. There seems little doubt that dramatic changes in our society will come about as worldwide communications become quicker, but people are suffering from data overload. There are many difficulties in distilling a small amount of useful information from vast quantities of data.

CHARTER 88 AND THE CAMPAIGN FOR A NEW UK CONSTITUTION

One of the trends in the UK related to changing community values is the growing demand for a written constitution, more democratic processes and a fairer deal for people in all walks of life. With this in mind a few people advertised Charter 88 as a reaction to the 'elective dictatorship' created during the years when Margaret Thatcher was Prime Minister. The initiative was launched to celebrate the 300th anniversary of the 1688 protest against Royal tyranny.

When Charter 88 was formed no organisation was intended – the aim was to collect signatures to support a published statement. However, the weight of popular support was overwhelming with strong demands for a campaigning organisation. Charter 88 is not a membership organisation. Those wishing to give support sign the statement of goals, a copy of which is given in Table 2.8.

Table 2.8 Charter 88 Goals

1   Enshrine, by means of a Bill of Rights, such civil liberties as the right to peaceful assembly, to freedom of association, to freedom from discrimination, to freedom from detention without trial, to trial by jury, to privacy and to freedom of expression.

2   Subject executive powers and prerogatives, by whomsoever exercised, to the rule of law.

3   Establish freedom of information and open government.

4   Create a fair electoral system of proportional representation.

5   Reform the upper house to establish a democratic, non-hereditary second chamber.

6   Place the executive under the power of democratically renewed parliament and all agencies of the state under the rule of law.

7   Ensure the independence of a reformed judiciary.

8   Provide legal remedies for all abuses of power by the state and the officials of central and local government.

9   Guarantee an equitable distribution of power between local, regional and national government.

10   Draw up a written constitution, anchored in the idea of universal citizenship, that incorporates these reforms.

Source Charter 88

Support for the campaign has grown steadily over the period 1988 to 1996, as shown in Table 2.9:[31]

Table 2.9 Membership of Charter 88

Year	1988	1989	1990	1991	1992	1993	1994	1995
Signatories	340	9,000	22,000	28,000	34,000	40,000	50,000	60,000

Charter 88 attracts a variable number of new signatories every month, depending on the nature and scale of activities around the country, but 1000 new supporters in one month is not uncommon. There are now 75 Groups throughout the UK. At least one in five Chartists (signatories) take an active part in campaigns such as letter writing, lobbying and organising Democracy Days. The success of Charter 88 has been reinforced in recent years by growing support from the main political parties and environmental groups like Friends of the Earth (FOE). This demonstrates the relevance of the aims of Charter 88 and recognition that it will help organisations like FOE to meet their own goals. Stronger democratic processes are believed by many to be a prerequisite of a sustainable society.

CONCLUSION

Higher material standards of living, greater social justice, improved education and more participation in decision making have been acheived in recent years. Despite this, many people continue to feel dissatisfied. Societies are more confused and fragmented and many people find it necessary to change some of their fundamental beliefs during their lives.

The single-minded pursuit of economic growth as measured by GNP is no longer universally accepted as the most appropriate measure of progress. Instead new measures, such as the Index of Sustainable Economic Welfare (ISEW), are emerging. They demonstrate that if GNP is adjusted for the damage it does to people and the environment as well as allowing for productive work that is not counted in GNP, a more realistic measure of progress is created. The ISEW suggests that quality of life has declined during the past 20 years while GNP has increased. Maybe this is why so many find the 'feel good factor' is illusory.

Some people are changing their behaviour by taking voluntarily action in ways which reflect their values, demonstrate their preferences and express their views. Traditional beliefs are giving way to emerging ideas and sowing the seeds of a new approach.

People with money to invest are turning to ethical and environmental investments and over £1 billion is now invested in this way in the UK. Energy efficiency and renewable energy are opening up new markets as their use increases. Householders are searching for new forms of financial help, are joining LETS and changing their lifestyles by participating in GAP. Diet and health concern many people and a growing minority are turning to vegetarianism and complementary medicine. The interest in organically grown food is increasing and more farmers are turning to this method of farming.

The Real World Coalition has brought together a wide variety of charities which are campaigning for realism in addressing common interests. Charter 88 is campaigning to achieve a written constitution for the UK to strengthen democracy and human rights. Other organisations believe their interests are more likely to be realised by working with Charter 88.

All these initiatives show that people are finding creative solutions which help them to improve their quality of life. Maybe these trends will continue to grow, but even if they remain as minority interests they could provide ideas which are helpful for sustainable societies.

There is scope for businesses of every kind to understand better what people are doing through their voluntary action. The ideas may not have immediate application for business but they may help businesses to understand better our changing world and stimulate interest in opportunities for:

- designing and developing new products and services;
- modifying trading practices to match human needs more closely; and
- enabling enterprise to become more ethical and profitable.

33

3 MARKETS
AND MINDSETS

The previous chapter concentrated on some of the ways in which change is occurring in the community and some changes that people are initiating for themselves. As yet these are small scale but may contain useful ideas for a sustainable future. In this chapter the focus shifts from the community to the business perspective. This includes changes in the market-place and how new thinking about business is beginning to emerge.

Numerous studies analyse changes in the market-place including environmental buying behaviour. In addition to research there are several topical ideas about how companies can respond to the environmental challenge, sometimes led by companies themselves. In addition to market surveys the mindsets of senior executives determine how companies deal with their social and environmental responsibility. Together the surveys and the mindsets provide the personal agenda for an increasing number of senior executives. These issues are explored using three broad themes:

- market surveys;
- mindsets which inform action; and
- the personal agenda of executives.

MARKET SURVEYS

Entec in association with The Green Alliance carried out a survey in April/May 1996 among the Top 1000 UK companies and 40 leading opinion formers on business and environmental issues.[1] The key findings from this survey include:

- environmental issues are moving higher up the corporate agenda;
- companies are concerned about safeguarding the health and safety of their employees and about water and air pollution, waste disposal and contaminated land;
- external pressure has stimulated a majority of companies to go beyond mere compliance; and
- many companies are not yet convinced of the business case for investing in the environment and main board support is still weak in many cases.

This survey adds to the MORI findings reported in Chapter 1. They show that top executives believe that not enough attention is paid to environmental matters despite being one of the most important factors in decision making. MORI research and the findings from the survey at the Strategic Planning Society conference both suggest that top executives, *as individuals*, are very concerned about environmental matters (see Chapter 1, page 11). However, there still appears to be some lack of conviction among boards of directors that more investment is justified. Some additional analysis and arguments for taking action are developed in this chapter.

TOMORROW'S COMPANY

The Royal Society for the Encouragement of the Arts, Manufactures & Commerce (RSA) brought together 25 of the UK's leading companies in January 1993 to establish a shared vision of Tomorrow's Company.[2] They published their interim report in February 1994 and the final report in 1995. The main conclusion is that to achieve sustainable success in an increasingly competitive market an inclusive approach is needed. An inclusive approach means working with and being responsible to all stakeholders – customers, suppliers, employees and investors. The report argues that this is the best way to discharge the traditional responsibility to shareholders. Added insight about the meaning of the word 'stakeholder' is given in Table 3.1.

Table 3.1 Who are Stakeholders?

An interesting definition suggests that stakeholders are those who become exposed to risk from the activities of a company either voluntarily or involuntarily, now or in the future. This definition enables a company to take a wide view of its responsibilities and establish appropriate management methods and feedback loops. It emphasises their ethical responsibility which goes wider than their legal obligations. It is a definition which can apply to public sector and voluntary organisations as well as businesses. The overall effect is to make all organisations responsible to people and the environment, in terms of their present and future liabilities. The purpose of enterprise in these circumstances would be to add value for all stakeholders by converting their stakes into goods or services. Although public bodies do not have customers they do have clients for their services. The definition and explanation of its implications adds depth to the inclusive approach idea.

Source Shann Turnbull, *New Economics*, Autumn 1996.[3]

The Tomorrow's Company Report puts forward a set of recommendations for action that can be taken by the government, institutions, companies and those involved in helping to bring about change. To aid the process of dissemination and implementation of the report's findings and recommendations the Centre for Tomorrow's Company has been formed.[4]

Successful companies of the future, according to the Tommorrow's Company report, will need to acheive world class standards, adopt the inclusive approach, behave responsibly towards the natural environment and seek business opportunities which are compatible with sustainable development. The readiness of customers and markets is an important consideration.

MORI MARKET RESEARCH[5]

MORI devised their concept of the 'socio-political activist' typology in the early 1970s. This identifies about 10 per cent of the British public as the 'movers and shakers' of British society. A similar approach was devised in 1988 (with slight revisions in more recent surveys) to describe the 'environmental activist'. A list of actions which people can take is used to find out what people say they have done in the previous year or two. The description of activities includes: walking in the countryside, donating money to wildlife and conservation organisations, selecting one product over another for environmental reasons, requesting information from an environmental organisation and supporting an environmental group through membership. Those who say they do five or more of the items on the list of fifteen activities are described as 'environmental activists'. The responses are not verified to see if people actually do as they say nor how frequently they do them. However, the method establishes a basis for comparison from one year to the next.

The responses to the question about 'buying one product over another because of its environmentally friendly packaging, formulation or advertising' is used to identify how many people can be classified as 'green consumers'.

MORI environmental surveys have been carried out in this way since 1988. The figures for environmental activists (EA) and green consumers (GC) are remarkably resilient. They are shown in Figure 3.1 as a percentage of respondents.

Figure 3.1 British Environmental Activists and Green Consumers

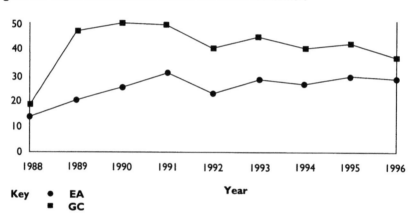

MORI also ask how strongly people feel about specific environmental matters by indicating if they agree or disagree with certain statements. The statements are shown in Table 3.2, together with the average percentage respondents who agreed or disagreed with each statement over the period 1993 to 1996.

Some people are cautious about accepting the MORI findings. Simon Bryceson, Director of Public Affairs, Burson Marsteller says:

if the Green Consumer existed on anything like the scale often claimed by politicians and environmentalists, there would be no need whatsoever for environmental legislation or regulation.[6]

There is some validity in this observation, however, it should be borne in mind that MORI research asks people to describe their own behaviour without independent confirmation and they do not ask people if they carry out their actions all the time. Even Bryceson suggests that people believe that environmental concern should be reflected in buying behaviour. MORI research surveys suggest that there would be even more consistency in the behaviour of people if fiscal measures encouraged green consumers, if there was less misleading advertising and clearer product labelling.

Table 3.2 Statements about the environment

Statements	Percentage
There isn't much that ordinary people can do to help protect the environment – those who DISAGREE	65%
Pollution and environmental damage are things that affect me in my day-to-day life – those who AGREE	70%
Too much fuss is made about the environment these days – those who DISAGREE	71%

MORI surveys find that 24 per cent of British adults claim to 'avoid using the services or products of a company which you consider has a poor environmental record'. This means that 10 million people in the UK seek to influence manufacturers through their buying behaviour, some of the time.

Questions are asked about which scientists are most trusted by the British public. As a result of events in 1996, over BSE and Brent Spar in particular, confidence in British government scientists has fallen. The public are also slightly more sceptical about the veracity of statements made by scientists employed by industry. Scientists employed by environmental groups are trusted significantly more than those employed by the government or by business, as shown in Table 3.3.

Table 3.3 Confidence in British Scientists

	Public confidence is Great/Fair	Public confidence is Great/Fair	Public confidence is Great/Fair
Scientists working for:	1993	1995	1996
Government	38	38	32
Business/Industry	41	48	45
Environmental Groups	73	82	75

The 1995 survey was carried out prior to the incident involving Shell's disposal of the Brent Spar. The 1996 survey was after Shell had reversed their decision to sink the Brent Spar in the Atlantic. MORI point out that when the phrase 'our scientists tell us...' is used by anyone, the public are more inclined to believe the environmentalists rather then the government or industry. Greenpeace miscalculated the amount of toxic waste in the Brent Spar and this has had some adverse effect on the reputation of environmental scientists. The long term effect is hard to assess. Greenpeace admitted their error quickly and frankly and this lessened the damage but they still angered other environmental groups.

MORI also explore the opinions of MPs. They find that eight in ten Opposition MPs, agree with the statement 'British companies do not pay enough attention to their treatment of the environment'. Furthermore, nine in ten Opposition MPs and a majority of the British public believe that the penalties for causing pollution are not severe enough. Both these findings could have considerable significance if government priorities change.

PRODUCT LABELLING

The Council of European Ministers reached agreement about the Eco-labelling Regulation in December 1991 including the use of an agreed logo. The scheme has two primary objectives:

- to inform customers about the environmental impact of products; and
- to promote and develop products with reduced environmental impact.

The product groups which were agreed for the launch included paper products, detergents, paints and soil improvers. For labelling purposes products are assessed in terms of all stages of their life cycle. This includes pre-production, production, distribution, the product in use and disposal after use. They are also assessed for the following forms of environmental impact:

- waste;
- soil pollution and degradation;
- air contamination;
- noise;
- consumption of energy;
- consumption of natural resources; and
- effects on ecosystem.

The process for obtaining an eco-label involves several stages, namely:

1 Application from manufacturer with relevant information.
2 Assessment of application and decision by country authority.
3 Member States advised and allowed 30 days to object.
4 If no objection the label is awarded and publicised.
5 If objections raised a further 15 days is allowed for resolution.
6 If unresolved referred to EU Regulatory Committee.

Once an application has been approved the manufacturer is then entitled to use the eco-label symbol on products and in advertising. However, since 1991 the whole process has proved to be more complex than envisaged and there are still very few approved eco-labels.

Eco-labelling is not the only approved labelling scheme. The Soil Association has its own symbol for approved organically grown food and farm/garden products. The Forest Stewardship Council (see Chapter 6) has its symbol for timber products which are grown using approved principles of forestry. The Marine Stewardship Council has embarked on a similar scheme in collaboration with Unilever. All these will help to provide customers with markings which guide people towards products which are compatible with sustainable development, but they need a lot of publicity to make them known and to build public confidence in the claims that are made.

Meanwhile customer confusion about product claims continues, scientific opinion about specific issues is sometimes changed or modified and there is little doubt that some customers have given up trying to find products and services which cause least damage to the environment.

SOCIAL VALUES

Changes in society have been tracked for at least 25 years; historians may say that they have always been tracked! The Central Statistical Office has published *Social Trends* each year since 1970 and its content has evolved over the years to reflect topical issues. For example there is now a section on environmental issues. These reports, however, tend to assume a 'business as usual' perspective. If a more radical approach is taken the government can get upset - especially if it is conservative and unwilling to countenance more radical indicators or more radical interpretation of published trends.

Francis Kinsman's book *Millennium*[7] draws on the work of *Applied Futures* [8] and merges this with his own experience. The research identifies three basic types of people in society:

- sustenance-driven, who are motivated by their need to survive and belong;
- outer-directed, who wish to succeed and show this with status symbols; and
- inner-directed, who are confident, seek self-fulfilment with wide interests.

In the late 1980s the research identified the three types of people who were fairly evenly balanced in Europe with 30–35 per cent of society in each category. The UK and the Netherlands had more inner-directed people (42 per cent) than other European countries, while Scandinavia had fewer sustenance-driven and Germany had a higher proportion of outer-directed people. During the 1990s there is some evidence to indicate that the three types have become more polarised as the tensions in society have increased. There are also signs that the inner-directed have divided more clearly into the 'carers' and the 'environmentalists'. The carers are particularly concerned about the state of society and the consequences for people, especially the poor, while the environmentalists are more concerned about the state of the natural environment, the consequences for its ecology, the biological diversity of our planet, our life support systems and the Earth's carrying capacity.

The survey conducted in 1994 by Synergy Consulting was designed to provide business clients with the ability to:

- understand the mindsets of people;
- assess the impact of social influences on individuals and groups;
- appreciate the dynamics of social and market changes; and
- evaluate the potential for future strategy.

Underlying this approach is a framework for understanding how values, beliefs and motivations influence attitudes and lifestyles which in turn influence how people behave. Values develop and change slowly, usually over 10 to 20 years. Attitudes and lifestyles can change within 2 to 4 years, and behaviour can change more quickly and easily – within a period of 0 to 2 years. The model describing this is shown in Figure 3.2.

Figure 3.2 Values, Lifestyles and Behaviour[9]

Values Beliefs Motivations Change takes place over: 10–20 years	➡	Attitudes Lifestyles 2–4 years	➡	Behaviour 0–2 years

The differences between the sustenance driven, outer-directed and inner-directed people is shown in Table 3.4 below.

Table 3.4 Comparative Values

	Sustenance Driven	**Outer Directed**	**Inner Directed**
Home	My home is simple – it's all I can afford.	My next home will be bigger & better.	I like it and it's comfortable.
Lifestyle	I struggle to maintain my standards, everything is so expensive.	I like to keep on improving my standards of living - that's progress.	My values guide what I do; expenditure is only one consideration.
Car	It often breaks down but I can't afford to replace it.	I like to have a smart, powerful, new car replaced every two years.	It's reliable, not too polluting and serves my purpose.
Food	Food is so expensive, I buy what I can afford.	I enjoy good food and drink and often eat out.	I eat a healthy diet and feel all the better for it.

Awareness of the values, attitudes and behaviour of these different groups of people provides a useful framework for planning:

- market research surveys;
- product features, design and packaging;
- the advertising approach;
- promotional schemes for products and services;
- the relevance of monetary incentives; and
- the sort of people most likely to influence buying decisions.

WHAT STIMULATES BUSINESSES TO TAKE ACTION ON ENVIRONMENTAL MATTERS?

Legislation has been the most significant reason why organisations have taken the environment seriously. No responsible business wants its activities to be seen as illegal. The law can now be invoked against directors and no one wants to be imprisoned or fined because their organisation has been found guilty of illegal practices. Oil and chemical businesses in particular have incurred heavy costs when mistakes and disasters occur such as the Sea Empress running aground off the Welsh coast (1996) or Union Carbide's disaster at Bhopal in India (1984). Chronic pollution is also a problem, such as the Mediterranean Sea, where 100 million people from 17 different countries live in the coastal areas.

The reputation of a business can be damaged or enhanced by the way in which environmental matters are tackled. Even a company with a positive environmental reputation built up over many years can suddenly find itself in an extremely difficult situation. There is no doubt that public opinion can be a powerful force for change. As ethical issues, the pressure for environmentally responsible behaviour and disclosure of information gain strength the more resources need to be allocated to reducing the risks, avoiding damage and finding solutions that make business sense.

Another significant reason for adopting an environmental policy is to save money by becoming more efficient. A great many businesses have found that they can save considerable sums by cutting out wasteful practices in the use of raw materials and energy. The resulting reduction in resource use through greater efficiency also cuts emissions and reduces environmental damage.[11]

Other reasons which play their part include the desire to enhance the image and reputation of the organisation, improve health and safety for both employees and the general public and meet the changing needs of customers.

Organisations like Rank Xerox, Electrolux and Procter & Gamble have found that they can improve their performance, strengthen their competitiveness and are better able to attract and keep good people.

Despite these evident benefits a great many organisations, especially the smaller ones, have not yet adopted any effective management methods to deal with their environmental impact or to recognise that there are business opportunities in helping others to face the environmental challenge.

WHY ORGANISATIONS AVOID THE ENVIRONMENT

Large organisations can easily give one or two senior people a special assignment to find out about environmental issues, explore how their business might be affected and develop ideas for taking actions that make business sense *and* help the environment. In fact large organisations use creative assignments such as this to stay at the leading edge of their industry, to provide development opportunities for key people and to stimulate imaginative ideas especially from respected mavericks who might otherwise leave and become successful entrepreneurs elsewhere.

In smaller organisations the best that can be done is to ask or instruct someone to look at this issue *in addition to his or her existing job*. This does not work well because any manager knows how easy it is for urgent work to take precedence over important work. Environmental issues are frequently important rather than urgent unless the organisation has fallen foul of the law or is in danger of doing so. The urgent work relating to today's immediate problems of manufacture, sales and budgets is often all pervasive.

A hypothetical situation, based on reality, provides a framework for appreciating the possible dynamics of decision making about environmental matters. A company might set a goal to reduce energy consumption by 30 per cent within two years. The forces for achieving this goal could be:

- complying with current and pending legislation;
- reducing energy consumption and saving money;
- making more efficient use of materials to improve reputation;
- matching competitors to avoid falling behind;
- encouraging employees who want to help create a cleaner environment; and
- knowing that environmental improvement makes sense.

The arguments look powerful so it is surprising that all organisations do not leap at the chance to improve their business, save money, enhance their reputation and improve the environment. There are other forces which resist change, typically these might be:

- there is no coordinated effort to achieve a goal like this;
- the costs of achieving the goal are believed to outweigh the benefits;
- if it makes business sense it would have been done long ago;
- employees do not know how they could contribute to the stated goal;
- time consuming changes are needed, but there is already too much to do; and
- the money is not available for capital expenditure.

This simple example illustrates the sort of dilemma faced by many organisations especially those that are smaller and find it difficult to get the issue properly examined. In many cases the resisting forces are based on unexamined assumptions. For example a cost/benefit analysis has probably not been worked out for a range of options. 'If it makes business sense it would have been done long ago' is a dubious, but prevalent assumption. Thorough energy efficiency studies are relatively recent undertakings in many organisations. Employees at *every* level can remain unaware of ways in which their behaviour can make a difference especially when this kind of training seems remote from the 'real priorities of the business'. The idea that it is all too difficult is a typical reaction when change is being introduced. Finally, there is often the belief that it is pointless trying to get capital expenditure approved 'they would never agree to spend that sort of money' − another assumption which is often untested.

These two sets of forces interact with each other as shown in Table 3.5.

Table 3.5 Helping and Hindering Forces Analysed

Goal: to reduce energy consumption by 30% within 2 years.

Helping Forces ➔	◀ Hindering Forces
• complying with current and impending legislation • reducing energy consumption and saving money • making more efficient use of materials to enhance reputation • matching competitors to avoid falling behind • encouraging employees who want to create a cleaner environment • knowing that environmental improvement makes sense	• there is no co-ordinated effort to achieve this goal • the costs probably outweigh the benefits • if it makes business sense it would have been done long ago • employees do not know how they could help achieve this goal • time consuming changes are needed but there is already too much to do • the money is not available for capital expenditure

The next step after this analysis is the most crucial. The typical desire is to increase the helping forces until they overcome the hindering forces. When this strategy is adopted it is surprising how frequently new hindering forces emerge and the situation stays unchanged. There is another strategy which concentrates on the hindering forces. The question to ask is 'which hindering forces if reduced or eliminated would make a real difference and bring about change?'

Three of the hindering forces could be much more closely examined:

• the costs of achieving the goal will outweigh the benefits;
• if it makes business sense it would have been done long ago; and
• it would involve massive, time-consuming changes.

Many firms have found that energy savings are a realistic, practical achievement and give a good return on investment. The following examples are amongst many that have been well documented:[12]

• Forte Plc saved £180,000 in 1993 by installing Combined Heat and Power (CHP) in 60 hotels, with a payback period of 30 to 36 months.
• Triplex Safety Glass Ltd, part of the Pilkington Group, find that monitoring energy use is the key to instant energy savings. First Energy was contracted with a flat fee to provide Triplex's energy requirements with maximum efficiency. Savings in excess of the contract were shared and 30 per cent savings were expected in the first year.
• Coats Viyella Plc saved £12,000 a year by installing a redesigned, efficient lighting system in their garment warehouse with a three year payback. They also saved £25,000 per year by extracting heat from cooling water with a heat exchanger which paid for itself in one year.

Once a robust financial case is made out it becomes realistic to implement the change goals. Then to make it work well other hindering forces can be tackled or might, in some cases, disappear. For example in the hypothetical case it is reasonable to assume that a coordinated effort would be developed to implement the proposals and people would be trained to understand how to play their part. The massive, time-consuming change may well prove to be nothing like that and the inability to find the capital could disappear either by arranging an energy management contract or by re-allocating capital expenditure because of the attractive returns.

MINDSETS

Mindsets develop as a result of education and experience. People adopt a mindset and form habits that guide their thinking and action. In a changing environment inflexible mindsets may perpetuate thoughts and actions that are no longer appropriate for new situations. Awareness of the mindset that is in use and the ability to change attitudes to match evolving situations is an important skill. Some potentially useful ideas are emerging to help guide and modify mindsets for the future. Several ideas are considered below.

Who Needs It?

The report entitled *Who Needs It?*[13] suggests a framework for analysing whether a product or service meets 'the need test'. This includes identifying the opportunities or threats each product or service is likely to face during the transition to sustainable societies. This transition is more likely to come about as a result of lifestyle changes rather than by cutting out specific forms of consumption. Lifestyle issues take into account values, visions and quality of life. The transition towards sustainability could shift the focus for business towards changing customer demand and is likely to require:

- environmentally responsible products and services;
- ethical business, including ethical and environmental investments;
- more disclosure of information about products and their performance;
- more efficiency in resource usage (for example 'more from less'); and
- better measurement of performance using a wider range of criteria.

Design Considerations

As environmental management becomes increasingly strategic the opportunities for designers to play a major role increases. However, a distinction needs to be made between the major design considerations involving products, choice of materials and packaging and detailed design which usually involves presentation of the product or service. Both are important but in this section the major design considerations provide the focus. The scope for encouraging new ideas has never been greater and smaller more agile organisations often have the advantage. Reviewing major design issues comes from imaginative and enthusiastic people and means looking at such things as:[14]

- new perspectives on old problems requiring imagination and innovation;
- radical new ideas which require new ways of thinking; and
- imagery, metaphors and visions which open up possibilities.

These thoughts can be applied to the design of buildings, products, processes and services. All these areas offer scope for radical changes as well as small but significant adjustments which can make a difference in terms of more efficient use of resources, eliminating pollution at source, being more compatible with the needs of people and being more aesthetically pleasing. The designer can explore new ideas, create new solutions, find novel ways to test prototypes, stimulate the interest of others, link robust design with artistic considerations, anticipate future needs, and work with others in new imaginative ways.

The shrewd designer will find ways to integrate personal values, social factors and ethical considerations with ecological concerns, while ensuring economic viability. No one believes that it is easy but two designers who command respect and have earned solid, practical reputations are Dorothy Mackenzie[15] and Victor Papanek.[16] Some examples of good design taking account of ecology, social factors, ethics and aesthetics are:

- a safe, movable playground designed so that children, teachers and parents, working in cooperation, could easily assemble it (Papanek p63);
- a simple trailer designed for collecting sorted garbage for recycling from a small community (Papanek p66);
- using sunlight creatively in building design provides heat and light in optimum ways (Mackenzie p42 and Papanek p79);
- the NMB Bank[17] and ING Bank,[18] both in The Netherlands have brought together human and environmental consideration in the design of their headquarters buildings; and
- the renewed interest in the bicycle is leading to the use of new materials and new technology to make this mode of transport even more useful and versatile (Mackenzie p79).

Developing Environmental Management

Some companies have been taking environmental issues seriously for many years, others have done so during the past five years, many more have recently begun to realise its importance but many are still inactive. Typically companies pass through four phases of environmental management, shown in Figure 3.3 as four steps.

Figure 3.3 Four Phases of Environmental Management

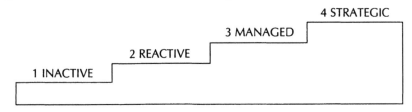

45

This framework can be developed to show the characteristics of the active phases. Three perspectives are used to describe these characteristics. The mindsets which are often prevalent, especially among top executives, the assumptions about the market which are either explicitly stated or implicit in how a company behaves. The third characteristic describes the management approach that is likely to prevail. These are summarised in Table 3.6.

Table 3.6 Phases of Environmental Management

Reactive

Mindset – business as usual: act on other issues if pressured;

Assumptions about the market – shows typical change pattern;

Management approach – current practices satisfactory with technical fix for unusual problems.

Managed

Mindset – cost-effective and efficient use of resources with some reduction in emissions;

Assumptions about the market – conventional market forces prevail but watch for emergence of changing customer preferences;

Management approach – a systematic approach introduced; using life cycle analysis, environmental audits and reports.

Strategic

Mindset – business strategy integrated with social and environmental factors; inclusive approach; use ethics and enlightened self-interest as basis for innovation;

Assumptions about the market – aware of need for long-life products, resources conservation, zero pollution to match customer needs;

Management approach – intergrated, holistic, systems approach with open reporting and independent verification, linked with financial performance.

The experience of businesses which have progressed through these stages is that the reactive approach becomes inadequate when the number and frequency of issues increases. Sometimes an event occurs which leads to significant changes in approach. Some examples are given in Appendix 1 to illustrate the different circumstances which stimulate organisations to change their mindset and take more dramatic action.

Most companies have found that in recent years the management of change is a continual challenge. Just as one issue is resolved a new topic arises. For example once waste has been reduced significantly the focus shifts to stopping pollution at source. This is partially achieved and the need for product re-design emerges. Before that has gone very far some social pressure indicates the need for further change. Eventually this process suggests the need to integrate social and environmental matters more closely with the mainstream business. Business strategy impacts on every issue and a more cohesive approach is required. A few companies that are doing this now are described in Chapter 5. These companies manage their social and environmental responsibility as part of their business strategy rather than alongside it.

When this happens environmental managers are involved in strategic thinking because they have earned top management respect for their competence. This is often achieved through implementing effective environmental management in ways which make business sense. To do this they have risen above tackling each issue separately in a reactive way and have developed a coherent, systematic approach. However it is not clear how environmental management will develop in the future. Environmental management is evolving and four scenarios are suggested in Table 3.7.

Table 3.7 Future Scenarios for Environmental Management

Option	Scenarios	Business response	Action and outcomes
A	Interest in environmental management declines.	Few companies have environmental departments; disbanded and very few new ones form.	Environmental management diminishes with a rise in wasteful and polluting practices.
B	Environmental management remains a discrete activity but gains strength.	Larger, more professional environmental departments are created to deal with more complex issues and regulations.	As complexity increases environmental management gains status and authority, requiring more qualified and influential people.
C	Environmental management is integrated with business strategy thus becoming part of line management responsibility.	The emphasis moves from improving efficiency and becomes part of business strategy with influence on product and process design. Managers are more directly involved. Environmental departments become more specialised.	Integration with business strategy (already happening in some companies) increases with support from small, specialised, professional environmental departments. Staff at all levels receive appropriate training and development.
D	Effective business leadership involves the inclusive approach and sustainable development.	A stakeholder approach is adopted, with explicit responsibility towards customers, employees, suppliers, shareholders, the community and the natural environment.	New measures of performance are developed and the goal becomes building to last to match stakeholders' interests. Manager and staff development, and a systemic approach reflect this change.

Option A seems unlikely considering the increasing political attention being given to environmental matters and the continuing number of environmental incidents. Option B reflects the growing strength of scientific evidence about environmental dangers such as global warming, health risks, the side effects of abusing nature and the frequency of major pollution disasters. However, this option relies on specialists.

Option C indicates that social and environmental issues move more strongly into the mainstream of business strategy. This might occur as a result of pressure from resource constraints (for example, food and water) and major fluctuation in prices (for example, petroleum and grain). Option D suggests that enlightened companies take a longer-term view of the evolving world situation and strive for a broadly based approach to their responsibilities and business strategy

Option A is not only consistent with companies that remain oblivious to the implications of environmental factors but suggests that the recent trend will go into reverse. Option B may be favoured by companies which struggle to keep environmental management at arms length from their main business strategy. Option C impacts on the design of products, modification of processes and the criteria for purchasing and procurement. It is improbable that changes of this kind will be undertaken by an environmental department without close collaboration with line managers. However, line managers will need to understand the issues, learn about ecology and appreciate how systems thinking can be applied in organisations in order to make an effective contribution. [20] Inevitably this will result in modifications to education programmes and the need for managers and employees at other levels to learn about environmental matters.

Option D takes matters much further and develops the ideas of Tomorrow's Company and an inclusive approach with sustainable development. While it is under consideration by several companies the form it will take is not yet clear.

PERSONAL AGENDA FOR ACTION

People from every section of the community, including many top executives working in the public and private sectors are concerned about society and the future. A significant number of these people admit that consumption is at times excessive, that environmental factors can be and are being taken into account when making buying decisions and most people will need to make radical changes to their lifestyles.

As more of these people make their views known and persuade their colleagues that radical change is essential a viable future for ourselves, our children and their children can be developed. This has global implications but requires action at every level – national, local and personal. Everyone can contribute to solutions if the will to do so is there. Personal example is always more impressive than public statements.

Papanek suggests three attitudes might form the mindsets of different people and he clearly recommends the third option:[21]

- each person tries to use less water and energy and re-use and recycle as much as possible;
- everyone leaves it all to the experts and to government; and
- **each person contributes to solutions from a personal perspective and role in society** [emphasis in the original].

CONCLUSION

Several market surveys show that public concern about environmental matters remains high. There is some evidence that senior executives not only share this concern but feel it more deeply than others. Several have been able to influence their companies to take effective action but some would like to see this taken further. An inclusive approach which acknowledges all stakeholders and safeguards the environment is being examined or adopted by a few companies.

Market trends and changing social values are important for all companies because they affect the way in which customers and potential customers react to products and advertising. MORI research findings and some social trends are described with indications of their relevance to all businesses.

Why companies take the environment seriously and why some remain inactive is explored. Mindsets are suggested as a significant factor in how people behave and some possible mindsets are described. Raising fundamental questions about products and services, such as *Who Needs It?* Is identified as a useful approach. There is also a brief overview of design considerations and approaches to environmental management.

In order to do justice to the scale of the challenge environmental management needs to be fully integrated with business strategy under line responsibility. The stages of environmental management are described. The strategic approach increases the importance of effective education of employees at all levels especially managers and senior executives. Some companies are already taking this further and integrating business strategy with social responsibility and environmental management. The inclusive approach is becoming reality for a few organisations.

Finally some options for a personal agenda for action are described. Everyone can play their part and individual contributions *do* make a difference.

4 SUSTAINABLE DEVELOPMENT AND
SUSTAINABLE ENTERPRISE[1]

Sustainable development, when applied to business focuses attention on the contribution that business can make to help create sustainable societies. This is a big subject that is complicated by various ways in which sustainability is perceived. Despite the difficulties, this chapter aims to clarify how businesses can approach the challenge and devise their own vision, objectives and approach. Relevant ideas are evolving, sometimes at a remarkable speed. In this situation values, ideas and processes are needed to encourage progress – broadly in the right direction.

It is highly improbable that sustainability is attainable in isolation by a single country, community or enterprise. The concept of sustainable enterprise is seen by some to be an oxymoron.[2] However, there are companies which were established over 100 years ago and many which would like to last at least that long into the future.

It is worth exploring a vision of sustainable society then considering how sustainable enterprise could both contribute to and be part of sustainable societies in an inter-dependent relationship. It is worth remembering that in nature sustainability is not achieved by any single plant having an endless life, nor even for all species to survive. A sustainable ecosystem includes several varieties of plants, animals and insects living together in an inter-dependent way, using nutrients from the soil, requiring water and relying on energy from the sun. How might something comparable be achieved in societies with help from business enterprise? Sustainable enterprise other than within the context of sustainable society seems implausible and sustainable societies are unlikely to be attainable without considerable help from enterprising initiatives in the private, public and voluntary sectors.

THE ROLE OF GOVERNMENT

The role of governments is crucial, but they alone cannot do all that is required – everyone should contribute. Government cannot leave it all to 'market forces'. The role of government is to provide the framework within which business enterprise and citizens can play their part effectively. Several national governments have done much to develop the approach to sustainable development, especially Canada, The Netherlands, Germany, India and the USA. The UK initiatives go back over 30 years and include:

- establishing the Countryside Commission in 1968;
- establishing the Royal Commission on Environmental Pollution in 1970;
- publishing *This Common Inheritance: Britain's Environmental Strategy* in 1990 and the subsequent follow-up reports;[3]
- establishing the Advisory Committee on Business and the Environment (ACBE) in May 1991[4];
- publishing *Sustainable Development: The UK Strategy* in 1994;
- publishing *Indicators of Sustainable Development for the United Kingdom* in 1996[5];

- establishing the Environment Agency, which published its Enforcement Policy in 1996; and
- setting targets for local authorities and encouraging them to achieve, as a first step, a reduction of household solid waste by 25 per cent by 2000.

Interspersed with these initiatives have been a host of laws designed to protect the environment and deal with some of the worst effects of unrestricted human activity. UK government thinking and approach has met with general approval but has been criticised by environmental groups for undue caution, compared with, for example, the Netherlands environmental strategy. Environmental groups and the Advisory Committee on Business and the Environment (ACBE) have expressed some disappointment with the speed of follow-up and implementation.

The areas in which governments have a crucial role to play are summarised in Table 4.1.

Table 4.1 Role of Government

Role of government	Action which helps the environment
Legislation	Developing and approving environmental laws.
Regulation	Establishing ways to ensure that legislation and regulations are implemented.
Fiscal measures	Using all forms of taxation and other fiscal measures to encourage voluntary actions that safeguard the living world. Adjustments in taxation can encourage conservation of resources and reduction in pollution.
Indicators	How to recognise whether progress is being made and to assess the health and resilience of the environment.
Advice and guidance	Setting up appropriate advisory services to aid progress towards sustainability.
Institutions	Ensuring that they are established and function in ways that support other measures, described above.

Ultimately, a sustainable society will have a stable population with lifestyles that can be accommodated by the Earth's carrying capacity. As people everywhere feel that their basic material needs are satisfied and they are better able to meet their aspirations, an underlying cause of conflict could be reduced. With pressure on resources eased there would be less risk of war, so peace between nations would be more attainable. By emphasising quality of life rather than the quantity of material consumption the willingness to close the gap between rich and poor could be more acceptable. However, all of this is far removed from the situation that prevails today.

SUSTAINABLE SOCIETIES

Sustainable development, is described as:

Meeting the needs of the present without compromising the ability of future generations to meet their own needs.[6]

The UK Round Table on Sustainable Development describes it as 'a continuous process – a journey, not a destination'[7]. However, they and many others seem reluctant to state the ultimate objective, which surely is to establish societies that are sustainable. Sustainable societies can be considered at many levels and contributions can be made by local communities, nation states and international organisations. Each community and each nation state is dependent on their neighbours also achieving sustainability. Individual countries can jeopardise not only their own future but the future of the world.

Striving for a sustainable global community means accepting a challenge that requires worldwide cooperation on a scale that has never before been achieved. And yet any other ultimate goal may well prove futile because the reality is that interdependence is a characteristic of life on Earth. Each country is increasingly capable of destroying not only the quality of life but the life support systems of the planet. This is a real threat rather than imaginary. It only requires one 'small' war to escalate into global warfare, for example this nearly happened in the Gulf War in 1991. Today, water shortages, world oil supplies, fishing in other countries' waters, and mass movements of refugees are examples of the triggers for serious conflicts.

These conclusions justify the search for a sustainable global society. A vision is offered as a possibility based on assembling the evidence of today's problems and pulling together the threads of a more healthy, more viable and more resilient society which stands a better chance of being sustainable. As new evidence emerges there is little doubt that a vision of this kind will be adjusted but it offers a start. A sustainable society needs to consider the perspectives, key elements and practical ideas set out in Table 4.2. A diagram incorporating the ideas set out in Table 4.2 is shown in Figure 4.1.

A radical shift in thinking will be required to achieve a society of this kind, as is neatly stated by Ervin Laszlo:[9]

It is often forgotten that not our world, but we human beings are the cause of our problems, and that only by redesigning our thinking and acting, not the world around us, can we solve them.

It is easy to dismiss this whole idea as Utopian and unrealistic but the alternative is to succumb to a future that is becoming increasingly untenable, demonstrably unsatisfactory and increasingly threatening to global security. Solutions are needed which offer a more satisfactory future and which can be accomodated within the Earth's carrying capacity.

Table 4.2 Vision of a Sustainable Society

Perspectives	Key elements	Practical considerations
Peaceful, democratic and satisfied communities	• Values	• Human rights
		• Justice and equity
		• Peace and security
	• Population	• Stable
		• Healthy
Economics as if people, environment and resources matter	• Government	• Democratic
		• Authoritative
		• Legitimate
	• Business	• Cyclic, conserving, healthy
		• Empowering
		• Profitable
		• Modified free markets
	• Technology	• Benign
		• Economical and efficient
		• Durable and appropriate
Conservation of physical resources	• Materials	• Re-used
		• Recycled and conserved
		• Sustainable sources
		• Used efficiently
	• Energy	• Renewable
Biodiversity respected and valued	• Water	• Clean water for all
		• Economical use
		• Pollution prevented
	• Ecology	• Life support systems safeguarded
		• Diversity of species protected
	• Food	• Sustainable yields
		• Near to markets
		• Organically grown

ENVIRONMENTAL IMPACT AND CARRYING CAPACITY

Paul Ehrlich devised the formula, $I = P \times A \times T$, where I is the impact on the environment, P is the size of population, A the level of affluence (broadly equivalent to the level of material consumption) and T the technology in use.[10] Ehrlich describes this as his I = PAT equation and it has proved to be an extremely helpful way to explore ideas about sustainability.

The equation has been adopted by the Real World Coalition[11] and modified to $I = P \times C \times T$ (where C = Consumption). The principle is the same but T is described as a measure of how efficiently the economy uses natural resources and produces waste. The equation enables the implications of reducing environmental impact in different circumstances to be explored. For example the following assumptions could be made for a given country:

- human impact on the environment – I – needs to be reduced by 50 per cent;
- population – P – is set to double in the next 50 years; and
- consumption – C – (that is the economy) grows by 2–3 per cent per annum.

Figure 4.1 Sustainable Society: A Vision[8]

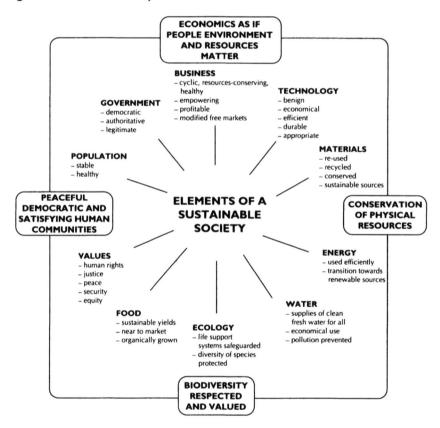

© Hutchinson, Colin *Vitality and Renewal* Adamantine, 1995. Developed from King and Schneider *The First Global Revolution* Simon and Schuster, 1991, The Problematique, pxi.

In these circumstances the efficiency of resource use –T– would have to come down to one-sixteenth its current level in order to achieve the goal of reducing environmental impact by half. That means improving the efficiency of resource use and reducing emissions by over 90 per cent.[12] The figures can be adjusted for different situations by changing the assumptions but there is little doubt that very dramatic improvements in efficiency are required in order to safeguard the environment – especially in industrialised countries which are responsible for the greatest impact on the Earth's life support systems. The impact is likely to be greater from the developing countries in the coming years because of their large populations and rapid rates of growth of their economies. If most of these countries rely on fossil fuels as the source of their energy the consequences for the whole world will be serious. However, renewable energy is being developed rapidly worldwide – see Table 2.3.

The carrying capacity of an environment or ecosystem is its ability to support life, adapt and renew its productivity. If an ecosystem is used regularly in excess of its carrying capacity it is damaged and the future carrying capacity is reduced.[13]

Further insights about carrying capacity are provided by the concept of an 'ecological footprint', with detailed calculations provided in a recent book.[14] Imagine a country like The Netherlands lifted off the surface of the Earth to form a bubble with an ankle and foot coming down to earth. In order to sustain the population with its current mix of lifestyles how much land would be needed to provide its resources and to absorb its wastes? The calculations suggest that this would be eleven times the land area of The Netherlands.

Alternatively consider Britain, which like other European countries, needs 3 hectares (7.4 acres) of ecologically productive land per person. The UK has an average of 0.35 hectares (0.9 acres) of ecologically productive land per person, so the deficit is 2.65 hectares (6.5 acres). Deficits for other European countries range from 2.15 hectares for Austria to 2.66 hectares per person for Germany. The USA, on the other hand, currently requires 5.10 hectares per person and has a deficit of 2.29 hectares. Japan and South Korea also have deficits but Canada has a surplus of 10.94 hectares per person.

An important point is that by living off 'nature's capital' it is feasible to exceed the carrying capacity for many years. However, the carrying capacity is then reduced and any subsequent sustainable society would have a much lower population. The same point is made in *Beyond the Limits*,[15] the sequel to *Limits to Growth*. The authors describe how an adjustment to qualitative development might be achieved:

A sustainable society would be interested in qualitative development, not physical expansion. It would use material growth as a considered tool, not as a perpetual mandate. It would be neither for nor against growth. Before this society would decide on any specific growth proposal, it would ask what the growth is for, and who would benefit, and what it would cost, and how long it would last, and whether it could be accommodated by the sources and sinks of the planet. A sustainable society would apply its values and its best knowledge of the earth's limits to choose only those kinds of growth that would actually serve social goals and enhance sustainability. And when any physical growth had accomplished its purposes, it would be brought to a stop.

TOWARDS SUSTAINABILITY

Three strategies are needed in industrialised countries to move towards a sustainable future:

- energy – improve efficiency in use of fossil fuels and convert as rapidly as possible to renewable forms of energy;
- develop long-life products and re-use materials, thus reducing demand for scarce resources; and
- drastically reduce emissions.

All three need to be accomplished in ways which acknowledge the importance of social justice, give priority attention to human rights and contribute to the relief of poverty.

In developing countries the priorities are different and might be:

- raising material standards of living, especially the relief of poverty;
- slowing down population growth;
- increasing use of energy, and accelerating the switch to renewable energy;
- transferring appropriate technology from industrialised countries;
- avoiding emissions at source; and
- improving social injustice.

People in developing countries wish to emulate the lifestyles of those in industrialised countries, while, at the same time, often seeking the freedom to do things in their own ways. There is no need for developing countries to copy the wasteful and polluting practices of the industrialised countries. These can be avoided by using more efficient processes, if technology transfer takes place with cooperation from developing countries. There are precedents. For example, African countries adopted motorised transport without the need for horse drawn vehicles. With goodwill, cooperation and technology transfer there is every reason to believe that developing countries could adopt modern technology without having to move through all the transitions that have occurred in industrialised countries. Of course this will not be easy and negotiations will be tough but it will be in the mutual interest of both developed and developing countries to work together on these issues.

At the present time developed countries tend to express concern about the rapid growth in population in developing countries whilst encouraging them to develop in traditional ways wherever this helps the economies of the industrialised countries. The developing countries criticise the developed countries because they have caused most of the damage to the world's ecosystems through their industrialisation and furthermore they set the ground rules for international trade which favours their own economies.[16]

Instead we need to bring about more mutual influence whereby:

- industrialised countries put their own house in order, create models which could help all countries and work out ways to transfer technology to developing countries;

- developing countries need to work out their own sustainable solutions, in their own interests, and influence developed countries to help this process by creating models which work and transferring efficient technology; and
- a healthy exchange of ideas and information is crucially important.

THE NATURAL STEP

The Natural Step is an imaginative approach developed in Sweden by Karl-Henrik Robèrt. Robèrt communicated with some 60 scientists of different disciplines in Sweden and achieved a consensus among them describing the environmental dilemma.

The essence of their thinking is that society cannot tolerate a systematic shift in environmental factors. There cannot, for example, be ever-higher levels of greenhouse gases and ever-smaller areas of agricultural land. The scientific foundation of the consensus can be summarised as follows:[17]

1 Matter and energy cannot appear or disappear (the conservation of matter and the first law of thermodynamics).
2 Matter and energy tend to spread spontaneously (the second law of thermodynamics).
3 Material quality is defined by concentration (purity) and structure. Energy or matter are not consumed, only their quality is changed.
4 Solar-driven processes create net gains of concentration and structure on Earth. Resource quality can only be maintained or improved by energy brought in from outside the biosphere. Photosynthesis is the only large-scale producer of material quality.

The Natural Step recognises that the prevailing 'linear' economies of the world are leading to the accumulation of thousands of substances by converting resources into waste faster than nature can cope. The alternative is to move steadily towards cyclic processes which conserve resources and are healthy. The scientific foundation coupled with the cyclic concept provide the basis for the four system conditions required for a sustainable society:

1 *Materials from the Earth's crust must not systematically increase in nature.* This means that fossil fuels, metals and minerals are extracted no faster than they are re-deposited into the Earth's crust. If this condition is not met quality is lost as wastes spread and accumulate towards unknown limits beyond which irreversible changes occur.
2 *Substances produced by society must not systematically increase in nature.* This means that substances are produced no faster than they are broken down and deposited into natural cycles. In practical terms this means planning for decreased production of natural substances that are accumulating and phasing out all persistent substances foreign to nature. If this condition is not met quality is lost as wastes spread and accumulate towards unknown limits beyond which irreversible changes occur.

3   *The physical basis for the productivity and diversity of nature must not be systematically diminished.* This means that 'harvests' of plants, animals and water must be within the capacity to regenerate and the extent of green surfaces is maintained.   Our health and prosperity depend on nature's ability to re-concentrate and re-structure wastes into resources. If this condition is not met then the Earth's carrying capacity is diminished.

4   *There must be fair and efficient use of resources to meet human needs everywhere.* This means that all communities need to use resources efficiently enough so that societies remain sufficiently stable and cooperative to succeed. This is needed to achieve the other three conditions.   In practice this means a more resource-economical lifestyle in developed countries and a levelling of population. Sweeping changes will be needed in land use, agriculture, forestry, fisheries, transport and urban planning. If this condition is not met social and economic breakdown of communities could lead to further destruction of the natural world.

The idea is spreading slowly to other countries including the USA and the UK. The Natural Step began as a grass roots initiative working in a low key way. The need to be economically viable and the importance of influencing opinion leaders has changed this to some degree.   However, The Natural Step does not provide answers but encourages individuals, professions and organisations to search for their own solutions in ways which make steady progress towards meeting the four system conditions.   The approach of The Natural Step and Forum for the Future, which is promulgating The Natural Step's ideas in the UK, are to commend examples of best practice and highlight achievements. Businesses looking for clear guidelines for their own business strategy are encouraged to explore answers to the following questions which relate to the four system conditions:

1   Does your organisation systematically decrease its economic dependence on underground metals, fossil fuels and minerals?
2   Does your organisation systematically decrease its economic dependence on long-lived unnatural substances?
3   Does your organisation systematically decrease its economic dependence on activities which encroach on productive parts of nature?
4   Does your organisation systematically decrease its economic dependence on using a large amount of resources in relation to added human value?

THE NEW AGENDA FOR BUSINESS

The new agenda for business is being defined by global organisations like the World Business Council for Sustainable Development and the World Business Academy. Both are challenging top executives and indeed people working at all levels to do some fundamental re-thinking about:

- the role of business in today's societies;
- the need for radical adjustment of core purpose and strategy;
- the sort of person who will succeed under new conditions.

The most powerful organisations in industrialised countries are large-scale businesses. They have the resources, the technology and the capability to bring about change; indeed they are already doing so. They are often criticised for the adverse effects of their activities and they are seldom praised for conserving resources, increasing their efficiency and creating products and services which lead towards sustainability. In many ways they bear the responsibility for showing that sustainable development is attainable. Stephan Schmidheiny has said: 'Sustainable development is a business issue that needs to be made a reality in each line function of every company.' (*Changing Course, p183*).

Guidance from Business Leaders

Guidance will need to come from enlightened top managers seeing the environment as a strategic business opportunity and not just as a corporate threat. This leadership can come from top executives of organisations in the public as well as the private and voluntary sectors, and can be greatly supported by the service sector and government action.

In order to accomplish this kind of transformation, fundamental re-thinking of how things are done will be needed. Companies are at the forefront of this challenge. Chief executives could provide more explicit leadership (a few are doing it already), while managers and staff at every level could re-examine their values and become increasingly involved in re-focusing the corporate mission. Goals and strategies to manage change more smoothly and in less time will be needed. This is the area in which organisations of every kind can strengthen their reputation, attract and keep good people and compete effectively in a changing world. Some are already well placed to demonstrate how authority can be delegated effectively, people trusted to behave responsibly and disagreements resolved amicably.

People responsible for the development of managers such as their managers, educators, trainers and some consultants have a unique contribution to make because they can enable the transition to be achieved more smoothly, aid the learning process and help those who are most severely challenged to come to terms with our changing world and its implications for them personally.[18]

Energy

One of the best indicators of material standards of living is energy consumption. To move towards a sustainable future requires richer countries to reduce their use of coal, oil and gas − initially by more efficient use − and to develop renewable sources of energy such as solar, wind, tidal, geothermal and biomass (dry matter and extracts from plants and animals) according to local circumstances and availability.

This is already happening and the generating capacity from sustainable energy alternatives (SEAs) such as wind power, solar power and fuel cells is promising.[19] The key points include:

- technological, demographic, environmental and other trends are converging to create a boom market for SEAs;
- worldwide wind power capacity increased by 35 per cent in 1995 and solar power, fuel cells, biomass, cogeneration and energy-from-waste also have impressive commercial potential;
- currently the primary markets for SEAs are the developing countries, followed by Europe; and
- demand in the United States is lagging but that is expected to change as the cost of SEAs declines and utilities commit to broad energy portfolios.

The generating capacity for wind energy is growing rapidly – see Figure 4.2.

In due course levels of consumption, in the richer countries will need to come down but quality of life can become a more important goal than continuously increasing material consumption. Recognising the distinction between material standard of living and quality of life is an important area for learning.

In the developing countries there needs to be considerable rise in material standards of living but, ideally, this needs to be coupled with efficient use of resources and steady development of renewable energy. Both industrialised and developing countries could learn from each other as they grapple with this challenge in their own ways.

Figure 4.2 Wind Generating Capacity by Region

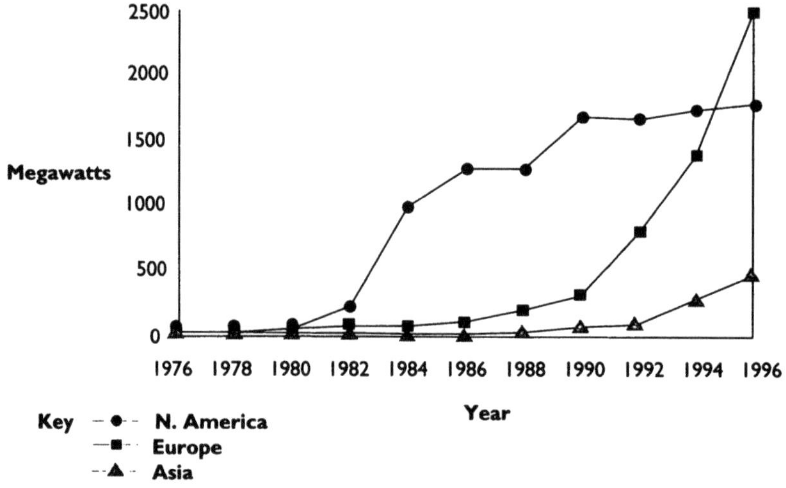

Key
- N. America
- Europe
- Asia

Transport

In addition to radical changes to the design of cars more use of public transport (which would need to be reliable, frequent, clean and efficient to attract customers) is important. It is even more important to attract freight back to rail, or better still find ways to reduce the distances that some goods are moved by purchasing locally. Better facilities for cyclists and walkers would help make both means of getting about more attractive for short journeys and could reduce automotive fuel consumption and the pollution that results from it.

Telecommunications can also help by replacing some travel and is much less extravagant on fuel consumption as described in Chapter 2 (p32). During the Gulf War many US companies grounded their staff, believing that flying could be too dangerous, but work still got done, suggesting that there may be scope to have fewer meetings and reduce business travel.

Reducing the energy used in transportation can be achieved by having fewer cars which travel four or five times further for each gallon of fuel. The Hypercar idea developed and publicised by the Rocky Mountain Institute (RMI) is an example of the sort of car that might be built in future.[20] Some ways in which Hypercars differ from the traditional car are outlined in Tables 4.3 and 4.4 below:

Table 4.3 Hypercars

Topic	Traditional	Hypercars
Design focus	Engine, transmission	Energy saving
Body	Metal	Carbon fibre
Frame/shape	Traditional	Infinite
Fuel	Petrol, diesel	Turbine with electric generator and photovoltaics
MPG	10–50	100–300
Gears	Gearbox	Electric motors
Steering	Wheels change direction	Selected wheels go faster
Fault finding	Garage	On board computer
Servicing	6000–12,000 miles	Like computer
Pollution	Gas emissions	Very little
Painting	Paint shops (capital cost often £250m)	Within moulded carbon fibre
Price	£6,000–£100,000c	£15,000–£30,000

A transition to cars of this kind will have considerable implications for different industries. Some will suffer disadvantages, for example steel manufacture, the oil industry and conventional motor manufacture. Others could reap massive benefits such as electronics, computers and designers.

The RMI could have kept the idea very quiet and found one manufacturer to develop the idea which could be a winner. If this approach had been adopted they would undoubtedly have stood a good chance of making a lot of money. However, they chose not to do that and instead made the idea available to anyone who was interested. The result is that some three years after the idea was first publicised there are now some 12 established car manufacturers and 12 in other industries, who could become Hypercar manufacturers, working on prototypes. Collectively they have committed about one billion dollars to support these developments and are competing to reach the market first. RMI claim to be 'using competition to get others to practice what we preach'![21]

Table 4.4 Contrasts: Conventional Cars and Hypercars

Topic	Traditional	Hypercars
Technologies	Steel manufacture, engineering, electronic	Petrochemicals, computers, electronic, mould makers, designers, entrpeneurs
Nature of Industry	Capital intensive, often lumbering	Vibrant, short-cycle, innovative
Countries of manufacture	Europe, USA, Japan	Brazil, China, India, Indonesia, Korea, Malaysia, Mexico, Nigeria, Taiwan – may be best placed for innovation
Disadvantages	Wasteful, polluting, sometimes unreliable, dangerous when badly driven	Not yet available. Carbon fibre bodies not recyclable. Does nothing for road congestion! Contribution to road safety unknown
Benefits	Mobility, convenience	Reliable, durable, comfortable, snappy, quiet, non-polluting, less prone to accidents, less vibration/noise

CONCLUSION

Progress towards sustainable enterprise requires international institutions, national governments, businesses and individuals to make their contribution. Key elements of the role of government have been described with examples of initiatives that have been and are being taken.

A model of a sustainable society has been suggested, while acknowledging that it may seem utopian and will surely be modified as more is learned about what is practical. The importance of carrying capacity has been emphasised and the concept of the ecological footprint has been described.

The work of The Natural Step has been summarised together with the non-negotiable systems conditions arising from the consensus they established among scientists in Sweden. Their work provides useful guidelines for business which offer challenging but clear ways for business to make real progress towards sustainability.

The emerging agenda for business, as stated by organisations like the World Business Council for Sustainable Development and the World Business Academy have been summarised and discussed under a few priority headings. The crucial role of business leaders has been emphasised, while acknowledging that many are already active.

People are the key to success and their willingness to accept even welcome change will be important. Radical new thinking will be required. Business has been transformed since the early days of the industrial revolution, but in many ways it is still often more like the voracious caterpillar before it becomes the beautiful butterfly. Contemplation of the butterfly and its earlier phases as a caterpillar and a pupa, provides a useful analogy for transforming business.

We should not regret the demise of the caterpillar but, instead contemplate the butterfly. By doing this we shift the focus from a voracious insect to a creature of beauty which treads lightly on the Earth, with great agility![22] The emergence of the butterfly is a natural example of transformation in practice.

Figure 4.3 Contemplate the Butterfly!

PART 2
THREE APPROACHES

The next three chapters look at initiatives that are contributing to the emergence of sustainable societies. There is still a long way to go but thinking and action is moving in the right direction. Examples illustrate the policy statements being made and how these are implemented. Obviously a comprehensive review is not possible in three chapters but a selection from large and small organisations describes what is happening. Professional bodies and local communities are also facing the challenge and some examples of their work are described.

Three Kinds of Contribution

Chapter 5 explores ways in which individual organisations are changing how they work, taking account of the environment, and integrating these changes into their business strategy. Examples from different sectors such as household, electronics, electrical equipment, banking, retailing and reclamation services are used. Other examples could be used but those selected illustrate sound principles that can be adapted by others. Both Electrolux and The Co-operative Bank provide good examples of organisations which are using the non-negotiable system conditions established by The Natural Step (see Chapter 4) to guide the development of their businesses. Examples from large businesses are easier to find but some smaller organisations are also taking action. Reclamation Services illustrate this point. There is considerable scope for many other new opportunities, especially from the 'sunrise industries' such as computing, communications, pollution monitoring, energy saving, renewable energy and waste management.

Chapter 6 looks at some interesting policy changes from industry associations and professional bodies. The Chemical Industry, Forestry, the Paper Industry, Marine Conservation, the International Hotels Environment Initiative and the Engineering Profession provide examples. They show how ethical principles, combined with enlightened self-interest can guide innovation towards solutions that enhance quality of life, make economic sense and reach out towards sustainability. Again, other examples could be used but the purpose is to indicate trends and encourage others to develop and publish their own contributions.

Chapter 7 examines how communities are creating visions of a more desirable future and then seeking ways to implement that vision with healthy practical outcomes. These initiatives typically involve a whole community including local residents, government officers, businesses, voluntary organisations, charities and other interested parties. Those most closely involved are often disappointed that there is not more willing cooperation from business. There is scope for worthwhile involvement from businesses of all kinds.

Kalundborg in Denmark shows how resources can be exchanged for mutual advantage. Kalundborg has a community-wide environmental committee to guide further progress. Curitiba in Brazil is a remarkable example of involving the community to make dramatic improvements to the quality of life of residents. Local government authorities are also doing interesting work as they move forward with Agenda 21. Industrial Ecology is emerging as a practical idea in Europe and America. All these provide heartening examples of what can be achieved.

5 ORGANISATIONS
GOING IT ALONE

Some companies have achieved a great deal by devising and implementing environmental policies. The pioneers deserve credit for the work they have done to reduce their environmental impact in ways that make business sense. The petrochemical industry led this initiative because of their considerable impact on the environment and the unfavourable public image of the industry. Their actions have resulted in a heartening reduction in environmental damage. However, the main reasons for taking action have been to comply with the law, improve safety and health, save money and enhance their reputation. Increased efficiency has been the main achievement but the easy pickings are no longer there for many large companies. The next stage will require a major change in approach and more far reaching improvements. This alone will bring about sustainability. It means integrating environment policy with business strategy in a systematic way.[1]

ENVIRONMENTAL POLICIES

The topics which have been included in environmental policy statements of major organisations for some years are listed in Table 5.1. Few organisations cover all these topics but the list is based on published statements by large organisations. Companies which are still confining their activities to this list are likely to be reacting to the environmental challenge. They may even be defensive in their approach.

Table 5.1 Topics Included in Environmental Policies

- Commitment to environmental goals.
- Compliance with and/or anticipation of the law.
- Periodic reviews and audits of operations.
- Publishing details of environmental impacts.
- Taking environmental factors into account in decision making.
- Applying the precautionary principle.
- Managing energy, resources and wastes more efficiently.
- Re-using and recycling materials.
- Purchasing policies adjusted for environmental factors.
- Raising standards for health and safety of employees and the community.
- Participating in community initiatives.
- Contributing to charitable causes.
- Training employees in environmental matters.
- Requiring suppliers to meet stated environmental standards.
- Reporting progress on achieving environmental goals.
- Using an environmental management system.

Among the companies which have made significant reductions in their environmental impact are Dow Chemicals, 3M and Monsanto Chemicals. All three have published details and have achieved a good return on the money they have invested in environmental management.[2]

Environmental reporting has become a new area for competiveness among larger organisations and awards are given for those whose reports are deemed to be the best.[3] This is a healthy trend but corporate environment reports (CER) can be improved in the following ways:

- they do not make close links with corporate financial performance;
- they tend to be kept sperate from annual reports;
- there is a danger of good CERs becoming an end in themselves for PR purposes;
- they are 'verified' or 'validated' in so many different ways that credibility cannot be assured;[4] and
- they do not address the issue of becoming a sustainable business.

FINANCIAL SECTOR

The World Business Council for Sustainable Development (WBCSD)[5] and the Advisory Council for Business and the Environment (ACBE)[6] are both calling for the links between financial performance and environmental performance to be made more explicit. The financial sector tends to ignore the CERs because they cannot see readily how environmental management affects financial performance. The WBCSD is urging the financial sector to support sustainable development. They recommend a thorough review of the criteria used for assessing the financial feasibility and viability of projects and proposals as indicated in Table 5.2.[7]

Table 5.2: WBCSD Critical Issues for the Financial Sector

- Sustainable development emphasises future values but they are discounted by financial markets.
- Sustainable development requires long payback periods, financial markets require short payback.
- Eco-efficiency often reduces current earnings in favour of future benefits but financial markets do the reverse.
- Sustainable development requires investment in developing countries but financial markets regard these as high risk.
- Existing accounting procedures do not recognise environmental costs which results in resources being wrongly allocated.
- The profitability of eco-efficient products and services is undermined by external costs being ignored in less efficient products and services.
- Fiscal systems usually favour labour productivity and neglect efficient use of resources.
- Environmental risks and opportunities are inadequately represented in current accounting and annual reports.

The WBCSD is undertaking a study to find out why financial goals are so often at odds with the goals of sustainable development. They wish to see the financial sector supporting corporate endeavour towards sustainability. The working group will publish reports in due course and these are likely to be included in future issues of *Tomorrow*.[8] Financial directors of all organisations could do well to make links with this study and explore the potential for asking these same questions within their own organisations.

COMPANIES AND SUSTAINABLE DEVELOPMENT

Some companies are already taking their environmental policies beyond the first four Cs – Compliance, Cost Savings, Care (that is safety and health), Credibility – and are reaching towards the fifth C – Customers (or Competitive advantage).[9]

When a company is content to work with these four Cs it may mean that it remains within the typical, conventional framework for doing business as shown in Figure 5.1.

Figure 5.1 Conventional Business

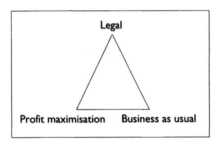

Many companies seem satisfied with staying within the law, striving for maximum profit and accepting business as usual as their operating philosophy. This approach may be legitimate if the product or service, as well as all the support services, do no damage to people or the environment.

Most businesses have been coping with continuous change for many years so 'business as usual' does not imply a static situation. However, for many organisations the core strategy has changed little. As the challenge of sustainable development is taken up the strategic options will need to be reviewed from many perspectives including sustainability. That could mean major changes for many industries.

A few businesses continue to market products which damage the environment or are manufactured by people working in poor conditions, earning extremely low wages. These businesses are failing to change with the times. It may be good for business in the short term but is demeaning and demoralising for increasing numbers of business people.

Several companies in different sectors have found novel solutions which break the mould of the conventional approach. The reasons why companies go beyond compliance depend on their individual circumstances. Examples of the triggers which have taken some companies more deeply into environmental matters is provided in

Appendix 1. These range from B&Q's fear of the unknown, through Monsanto's desire to pre-empt strict US legislation and turn a problem into an opportunity, to Norsk Hydro who were exposed by environmentalists and then became a leading exponent of responsible environmental management. By examining examples such as these it is possible to identify three factors which form the basis for integrating environmental and social responsibility with business strategy. They are:

- the *unique circumstances* of the specific organisation;
- the interest and commitment of an *influential champion*; and
- the *vision* of a more desirable future.

Another way to look at the potential opportunity has been described in terms of an ethical approach combined with enlightened self-interest. Chris Marsden describes the three elements of corporate social responsibility as:[10]

- to run your business effectively, profitably, safely, legally and ethically;
- to minimise the adverse impact of your activities on your social and physical environment; and
- to address those social issues which impact your activity, are within your competence to influence and offer opportunities for mutual benefit.

Marsden goes on to describe a process for taking action, how to distinguish between philanthropic action and profitable, responsible business practice and how to involve employees in the process. Masden's views can be summarised as shown in Figure 5.2.

Figure 5.2 Ethical, Profitable Business

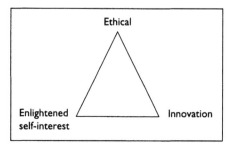

In this diagram *ethical business* combined with *enlightened self-interest* broadly cover Marsden's three criteria. However, ethical business and enlightened self-interest alone may not lead to effective action. The extra ingredient is *innovation* – a crucially important addition. The focus on innovation which is ethically sound and meets the needs of the business through enlightened self-interest, provides the dynamic which can lead to changes which meet social and ecological requirements and are profitable. For example we can look at the manufacture of CFCs. It is ethically right to cease manufacture, it is an example of enlightened self-interest and the scene is set for innovations which match these two criteria. This decision is simplified because it is now illegal to continue manufacturing CFCs. Some companies' vision goes beyond any legal requirement and includes clear messages about an ethical approach coupled with enlightened self-interest as a basis for innovation. For example, Edgar Woolard,

Chairman of EI du Pont de Nemours, says:[11]

If we are some day to have a truly sustainable world economy, we must begin to imagine a world chemical industry less dependent on petroleum. That could have broad implications for research, for the future of materials science, and for recycling and related technologies. In a truly sustainable economy, an energy company would seek to satisfy energy needs through a combination of renewable and non-renewable resources and technologies. Energy development would mean more than seeking new sources and markets for traditional fuels.

Figure X

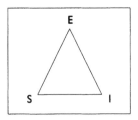

A shorthand version of Figure 5.2 (as shown above) will be used in subsequent chapters to draw attention to instances when it appears that the ESI (E = Ethical, S = Enlightened self-interest, I = Innovation) principles have been applied with success.

Some businesses are strengthening their performance and at the same time adopting an ethical approach which helps society and the environment through enlightened self-interest and innovation. Six examples shown in Table 5.3, have been chosen, out of many, to represent different industries.

Table 5.3 Examples of Progressive Companies[12]

Industry	Company Name	Turnover	Pre-tax Profit	Employees
Health and household	Procter & Gamble	£35.3 billion	£4.8 billion	103,000
Electronics	Rank Xerox Ltd	£4.1 billion	£570 million	22,000
Electrical equipment	Electrolux	£10 billion	£500 million	112,000
Financial services	Cooperative Bank	Not applicable	£37 million	3790
Home improvements	B&Q plc	£1.2 billion	£68 million	18,000
Salvage	Reclamation Services Ltd	£750,000	Not disclosed	12

PROCTER & GAMBLE

Procter & Gamble (P&G) celebrated their 150th anniversary in 1986 which gave John G Smale, former Chief Executive of P&G the opportunity to look forward to the next 150 years and say:[13] 'Our commitment must be to continue the vitality of this company .. so that this company, this institution, will last through another 150 years'.

A claim of that kind in a changing world establishes a remarkable challenge for all employees. However, P&G pride themselves on a long history of striving for excellence in their products, their honest and fair trading practices, continuous self-improvement and respect and concern for the individual. They positively relish challenging goals and say: 'we like to try the impractical and impossible and prove it to be practical and possible'.

Environmental issues first began to emerge in P&G in the 1950s and were at the forefront of business strategy long before they became relevant for many other companies. P&G is a good example of a company using the virtuous triangle described in Figure 5.2. P&G Environmental Policy affirms their:[14]

- commitment to operating in an environmentally responsible manner;
- continuously improving their products, packages and processes; and
- establishing innovative technology, processes and communications.

They have established their own way of integrating environmental factors into their business decision making through three specialist departments, a corporate-wide coordinating group and a unifying philosophy:

Professional and Regulatory Services makes sure that products are safe for people and the environment they disseminate research findings to scientists, medical profession, government and universities.

Environmental Science Department conducts basic research to establish how products and packages affect the environment.

Product Supply Environmental Department ensures P&G is consistent worldwide in measuring environmental performance and meeting regulatory requirements.

Environmental Quality Control Group coordinates the P&G worldwide corporate environmental programme and helps identify improvement opportunities across business sectors.

Environmental Stewardship provides the corner stone for thinking and action throughout the business on environmental issues.

The P&G Environmental Management Framework is described by Peter White in a paper presented to the Royal Academy of Engineers.[15] P&G's Environmental Stewardship approach is shown in Figure 5.3.

Figure 5.3 P&G Environmental Stewardship

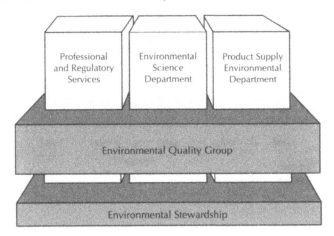

Environmental Management Structure depicted above ensures that environmental considerations are built into every aspect of the business. Such integreation is vital to the overall environmental performance.

The P & G philosophy empowers people to be innovative in the development of ideas and action including cross-fertilisation of best practice between all sites worldwide. Long-term thinking is encouraged and even when progress is accomplished in small steps the cumulative benefits are impressive. P&G carry these principles forward when working with others as they often do when tackling environmental problems in a societal context.

The areas in which improvements have been accomplished through the application of P&G's long-standing slogan *more from less*, include:

- Product design ⟶ re-thinking formulation
- Raw materials ⟶ using less
- Manufacturing ⟶ using less water and energy
- Packaging ⟶ smaller packaging
- Transportation ⟶ less energy for smaller, compact products
- Consumer use ⟶ smaller dosage

The crucial aspect of this work is that consumers are given greater performance and value ('more') while the environmental effects are reduced. 'More from less' can be applied by designers, manufacturers, waste managers and policy makers to help create sustainable product life cycles.

In 1994 the USA Superfund Act (SARA) required companies to report releases annually of more than 300 chemicals (rising to 600 from 1995). Since 1991 these releases have been reduced from 18.9 million to 6.1 million pounds (8.6 to 2.8 kilotonnes) and P&G is on track to achieve their targets as shown in Figure 5.4.

Figure 5.4 P&G's SARA Releases in the USA (313 chemicals)

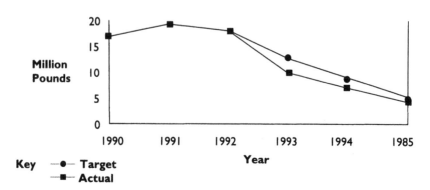

Equally impressive results are being achieved with the use of post-consumer recycled content (PCR) of major packaging materials as shown in Figure 5.5.

Figure 5.5 P&G's PCR Usage for Major Packaging Materials

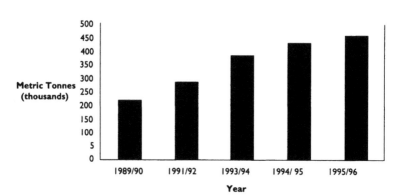

Among the achievements which P&G claim are the following:

- integrating environmental policies with business strategy so that they impact directly on product design, manufacturing processes and purchasing decisions;
- being the first company to bring recycled plastic bottles and refill packs to the laundry detergent market which helped to eliminate 80 million cartons per year;
- getting consumers, manufacturers, retailers, environmentalists and government officials at all levels to work together to seek environmental solutions;
- recognising that consumer decisions are based primarily on performance but environmental attributes of products are also important and customers need accurate, science-based information;
- publicly declaring the environmental targets, measuring performance against these targets and publishing the results; and

- over the period 1984 to 1994 P&G reduced the energy usage per production unit by 60 per cent by improving the efficiency of processes.

RANK XEROX

For many years Rank Xerox (RX) leased all their copiers to their clients. This created a disproportionately large asset on their balance sheet relative to their revenue. It became important to consider how to make more use of these assets. They explored ways in which to re-use and recycle copiers and their components and in due course started to manufacture them to make re-use and recycling easier. Today they sell copiers as well as leasing them and have broadened the base of their business to become 'The Document Company'. Paul Allaire, President and Chief Executive of Xerox Corporation says: 'Our environment policy applies to every aspect of our business operations – *including most emphatically the way we design our products.*'[16] [author's emphasis].

Their environmental report states:

*Protecting Planet Earth begins with a commitment which extends from products and services to where our customers work and live. **Environmental health and safety concerns take priority over economic considerations**.*

RX goals include product design, minimisation of waste, use of recycled materials and returning components to productive use after their initial lifecycle. Figure 5.6 illustrates the basic principles of Rank Xerox environment product stewardship and Table 5.4 provides explanatory notes to support the diagram.[17]

Figure 5.6 Rank Xerox Environment Product Stewardship

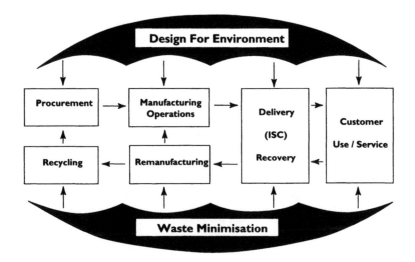

Table 5.4 Rank Xerox Product Stewardship Notes

Design for the Environment embraces the whole life cycle of products from the design and procurement of parts to end-of-life reprocessing and ensures that an increasing number of equipment components are re-used by being interchangeable and compatible across the different ranges of products.

The Procurement Vision is to work with suppliers to achieve full reprocessing capability of equipment parts and components.

Manufacturing Operations incorporate programmes for energy efficiency and reducing waste to a minimum. Where they have not been eliminated the use of harmful materials is being minimised.

The Integrated Supply Chain (ISC) provides a reliable return process closing the loop for the whole product life cycle. The 'Just in Time' approach reduces the need for storage and avoids waste. End-of-life equipment is returned to one of the RX Asset Recovery Operations (ARO) for further reprocessing. Machine assembly and configuration takes place at the European Logistics Centre to enable 'plug and play' on customer premises.

Customer Operations provide a key focus for RX product stewardship and customer requirements are monitored continually to ensure that operational, environmental and ergonomic needs are responded to. A wide choice of environmental papers are available and take-back schemes recover end-of-life equipment, consumables and used packaging.

Remanufacturing Operations and newly manufactured machines use the same people, assembly lines and testing equipment and carry the same performance guarantees.

Recycling backs up the re-use of parts and components so that as much residue as possible is recycled.

Waste Minimisation and Energy Savings are achieved through the '3-R policy' – Reduce, Re-use, Recycle. RX strives to minimise waste to landfill which helps prevent pollution and resource conservation and is highly cost-effective.

As a result of applying these policies Rank Xerox reduced its general waste (mixed packaging from suppliers accounts for 30 per cent) from 2300 tonnes at the end of 1992 to about 800 tonnes a year later and since then the figure has fluctuated between 800 and 1050 tonnes. The analysis of the electronics manufacturing waste stream shows several items are re-used (for example electrical components and recycled assemblies) or recycled (for example cardboard, polystyrene, paper, plastic cups and aluminium cans) but some items are sent to landfill and the quantities have not yet been measured (for example plastic reels, canteen waste, plastic trays and fibre glass dust). The percentage of the waste stream that is re-used is given in Figure 5.7.

Figure 5.7 Rank Xerox Waste Utilisation

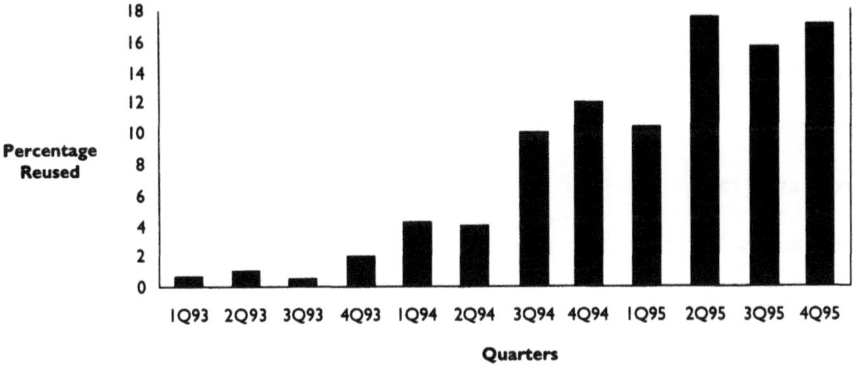

Other information and charts in the full report confirm several additional achievements:

- re-use and recycling has made the Product Life Cycle a closed loop;
- components reprocessed through Asset Recovery Operations were fewer than 100,000 in 1988 and had risen to 3,500,000 by the end of 1995;
- the quantity of machines manufactured through Asset Recovery Operations rose from 14,000 in 1989 to 60,000 in 1995;
- waste sent to landfill in 1991 weighed over 2500 tonnes per quarter but by 1995 this had reduced to just over 1000 tonnes per quarter;
- recycled materials represented 55.9 per cent in 1993 but this had risen to 87.6 per cent by the first quarter of 1996;
- environment is fully incorporated into the RX communications strategy; and
- RX have won several awards for their environmental work.

Pierre van Coppernolle, Director, Rank Xerox 2000 Strategy and Environment states the following in his concluding remarks at the end of their Environmental Performance Report, November 1995.

Our management teams are totally committed to the concept of sustainable development and we believe that Rank Xerox is amongst those leading companies which take the trouble to investigate continuously their environmental performance and then start to make improvements through pragmatic action. . . Our goal for the Waste Free Office campaign is to develop a showcase to customers by demonstrating that the responsible use of equipment and proper management of waste is not just the domain of The Document Company but an initiative that can be adopted by all companies that use The Document.

(pp38–9)

Rank Xerox Limited business results are given in Table 5.5.

Table 5.5 Rank Xerox Limited

Figures in £ million	1991	1992	1993	1994	1995
Turnover	2766	2923	3161	3309	4139
Profit after tax	117	81	110	100	332
Fixed assets	726	914	1020	1028	1090
Total capital employed	1371	1567	1640	1730	2039

ELECTROLUX

Electrolux is a Swedish company with worldwide operations. They have some impressive achievements as a successful business as well as with their environmental programme and now work with the Natural Step (see Chapter 4, page 57). Leif Johansson, Chief Executive and President of Electrolux, made a powerful statement in their 1994 environmental report:[18]

Today the industrialised world is faced with a number of challenges related to the environment and survival. For our part at Electrolux it means changes which will have an impact upon the entire company at all levels. In our efforts to satisfy market demands, we assume responsibility to contribute to sustainable development from an environmental perspective ... This is a strategic decision given high priority ... I am convinced that we are seeing the birth of a new perspective of the world, where ecology and economics are two sides of the same coin. This means we must strive towards greener solutions for environmental reasons, and also because it's economically profitable and good for us as a Group.

(p8)

Electrolux have a vision which recognises that long-term survival for the individual, for corporations and society means that there must be regard to the limits that nature can accept in the form of resource consumption and pollution. They recognise that growth in consumption of non-renewable raw materials cannot continue indefinitely. They state that their operations and products must be integrated in a cycle, so that customer needs can be satisfied without jeopardising the prospects for future generations. Their environmental policy is summarised in Table 5.6.

Table 5.6 Electrolux Environmental Policy

Responsibility to fulfil the demands for their products while contributing to sustainable development.

Precaution guides all development and production in order to avoid irrecoverable environmental impact.

The Total Approach from raw materials and production to utilisation and recycling involves life cycle analysis and helps focus attention on important issues.

Preparedness means acquiring new knowledge and readiness to respond in ways which meet future environmental needs.

Priorities are set to address defined goals in cost-effective ways.

Market leader credentials were earned when Electrolux was the first to introduce CFC-free refrigerators and freezers to the European market.

Profitability: effective use of resources is a decisive criterion for profitability and a prerequisite for environmental activities.

Electrolux endorse the consensus of scientists achieved by The Natural Step and state in their report:

We don't need to guess what nature's limits are. Scientific consensus has [established] the conditions of the ecological system. This frame determines the absolute boundaries for civilisation and every living thing. And it is not negotiable.

(ibid p8)

The systems conditions written into the Electrolux Environmental Report are the same as those included in Chapter 4 and they provide the central guiding principles for both business strategy and environmental protection in Electrolux. The Natural Step has devised a diagram to illustrate how an industry adopting a defensive strategy and striving to break the system conditions, could have severe fluctuations in its profitability. By contrast another company, such as Electrolux, which accepts the system conditions could work towards a more consistent approach with more predictable profitability. This is shown in Figure 5.8.

Figure 5.8 Profitability of Strategic and Defensive Companies

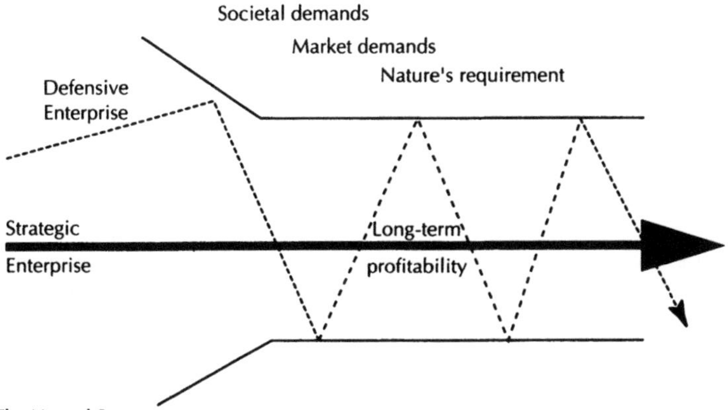

Source: The Natural Step

In their 1994 report Electrolux describe their approach and quote examples to show how they have made progress in meeting the non-negotiable conditions stipulated by

The Natural Step consensus of scientists. These include:

To reduce economic dependence on substances from the Earth's crust:
- traditional oil for chain-saws have been replaced by vegetable oils;
- metals such as copper and steel are recycled;
- railways are used for 75 per cent of product transportation within Europe instead of road transport; and
- the Husqvarna lawnmower uses solar cells for power rather than petroleum based fuel.

To reduce economic dependence on unnatural substances:
- brush-cutters and trimmers are fitted with catalytic converters and use 30–35 per cent less fuel, while emitting less hydrocarbon and nitrous-oxides (supporting systems condition 1);
- CFCs have been phased out from all European white goods and in 1993 a series of CFC-free refrigerators and freezers were launched; and
- a new generation of ploughs introduces a new system of cultivation which drastically reduces fertiliser usage and waste compared with conventional methods.

To minimise exhausting nature's biological diversity:
- the Frigidaire Company in Iowa, USA has reduced its water consumption and their water bill by 30 per cent by using a closed loop water cooling system; and
- Electrolux subsidiaries have contributed considerable sums of money in the USA, Canada and Italy to partnership projects with National Parks and the Worldwide Fund for Nature.

Resource wastage has been cut to get more from less by:
- more sensitive electronic controls systems regulate temperature in cooking appliances, use energy more efficiently and utilise leftover heat;
- the amount of material used in packaging is being reduced and higher proportions of packaging are being re-used through cooperation between Electrolux and its suppliers; and
- washing machines are now more efficient in use of both energy and water.

In 1995 Electrolux modified their reporting methods.[19] The Natural Step's system conditions are central to the development of business strategy, but not necessarily the best way to report achievements. In the 1995 report one of the most impressive achievements has been the reduction in the use of substances which have an Ozone Depleting Potential (ODP) and Greenhouse Warming Potential (GWP). Electrolux reduction in use of ODP and GWP substances in Europe are shown in Table 5.7. A new eco-labelling regulation in Europe requires refrigerators, refrigerator/freezers and freezers to be labelled ranking their energy efficiency in seven categories. Electrolux has about 500 models in the top two classifications and sell some 4.8 million of these products in Europe annually.

Table 5.7 Electrolux: Phasing Out Substances which have Ozone Depleting or Global Warming Potential (European figures).

	1992	1993	1994	1995	1996
ODP - thousand tons of CFC11 equivalent	3360	2520	240	0	0
GWP - Carbon dioxide in million kilograms	13584	10766	608	453	390

For the future Electrolux are ensuring that environmental thinking permeates all levels of their organisation. One of the priorities is achieving integration of this thinking across all business functions and with their suppliers. In the coming years they will:

- check performance against stated goals;
- implement a comprehensive change programme which can envisage and identify business opportunities compatible with environmental management;
- strengthen the links between group and business unit strategies; and
- find improved ways of reporting outcomes so that ecological and economic performance is more closely linked.

THE CO-OPERATIVE BANK

From its origins The Co-operative Bank rated social responsibility and participative management highly, so it was relatively easy to declare an ethical policy in 1992.[20] In 1996 the Bank named seven 'stakeholders' to whom they acknowledge a responsibility, namely: their shareholder, customers, staff and their families, suppliers, communities, society as a whole and past and future generations.[21]

Their Ethical Policy, which was developed with the involvement and support of their customers, states that they will not invest in or supply financial services to any regime or organisation which:

- oppresses the human spirit or denies human rights;
- manufactures or sells weapons to countries with oppressive regimes;
- is involved in speculation against the pound, money laundering, drug trafficking or tax evasion;
- manufactures tobacco products;
- exploits factory farming methods; and
- is involved in producing animal fur or in blood sports.

By contrast they will:

- actively seek customers who promote fair trade;
- encourage customers to be proactive about the environment;
- do business with those with complementary ethical values; and
- continue to strengthen their customer charter.

Results since 1991 have improved steadily as the figures in Table 5.8 show.

Table 5.8 Co-operative Bank Performance

Figures in £ million	1991	1992	1993	1994	1995
Pre-tax profit	-5	9.5	17	28	37
Operating income	195	209	228	238	249
Debt provision	48	42	38	32	22
Retail deposits	1050	1100	1310	1550	1980
Retail lending	1680	1480	1430	1500	1610

The Co-operative Bank has declared an Ecological Mission Statement based on the non-negotiable systems conditions established by The Natural Step (see Chapter 4). The Bank acknowledges that the social and ecological problems we face offer challenges and opportunities. It recognises the problems this creates for business especially the smaller business. The pressures on business include government regulations, lenders becoming more cautious, the need for energy efficiency and waste minimisation and the developing market for pollution control goods and services. An Ecology Unit has been established within the Corporate and Commercial Division. Its role includes making sure that the Bank's lending portfolio increasingly recognises the limits of the natural world to provide resources and absorb wastes. This is becoming a major criterion in assessing the risks and wisdom of alternative projects.

By establishing The Co-operative Bank National Centre for Business and the Environment, with four Manchester universities, a major resource to help businesses, voluntary organisations and the public sector has been created. Terry Thomas, Managing Director of The Co-operative Bank, has stated his intention to apply pressure on his colleagues in the financial sector having proved that ethical and environmentally responsible business is profitable.

B&Q PLC

B&Q started their environmental programme in 1990.[22] Alan Knight, B&Q's Quality and Environment Controller, whose interview with Kim Loughran was published in *Tomorrow*,[23] stated that a significant reason for getting involved with the environment was 'the fear of the unknown'. B&Q were closely questioned by the media about the origin of the timber they sold in their DIY stores in the UK. Despite the recession B&Q recognised that their customers, employees, local communities and the government all believed that the environment mattered. Jim Hodkinson, Chairman and Chief Executive says:

We've found in certain circumstances that by improving environmental performance we have made significant cost savings or have improved efficiency ... Through our environmental policy we have been recognised as a responsible retailer and there is no doubt that our corporate image has benefited. We have always said that our environmental policy must be a long term one.

(How Green is my Back Door? B&Q's Second Environmental Review, p3)

B&Q published their first environmental review in 1993, their second in 1995 and their third review is due in 1997. They have merged environment with quality – a strategy being followed successfully by an increasing number of enlightened companies. A retail DIY business like B&Q inevitably has a significant environmental impact and a high profile because they sell 40,000 products in 280 stores and employ 18,000 people.

When the environmental work began one of the most difficult questions to answer was 'where does the timber you sell come from?' Timber is not the only 'difficult' product. They are the largest UK retailer of paints and very aware of many potential problems. A third area of concern is purchasing from developing countries where a host of questions arise such as fair trading practices, employment conditions and the integrity of the sources of supply. B&Q, like many other companies did not know how to answer questions they were being asked. Rather than overreact B&Q put together an action plan with targets which was approved by the Board.

Table 5.9 B&Q's Timber Usage

Description	1993	1994
Total timber usage (cubic metres)	259,000	283,268
Volume of timber traced to country only (percentage)	1.41%	0.9%
Volume of timber traced to region within country (percentage)	0.23%	0.9%
Volume of timber traced to named processing mill (percentage)	59%	45.6%
Volume of timber traced to specific forests (percentage)	39%	52.6%
Number of countries supplying timber	41	50
Timber originating in the UK (percentage)	52%	54%
Timber from temperate forests	92%	91.6%
Commonest species used (Conifers – percentage)	84%	85.5%
Commonest tropical timber used (Rubberwood – percentage)	1.9%	2.1%

In a video which B&Q have produced they show two examples of their environmental work.[24] The first film describes a project in Papua New Guinea where villages no longer need to sell forest blocks at meagre prices to timber companies. This often results in ten trees being damaged for every useful tree extracted and also destroys the ecology of the forest. By training the villagers to select individual mature trees and using a portable saw mill, the timber companies are no longer needed. The trees are prepared by a locally established timber project and the villages receive 20 to 30 times the money they used to get. Independent certification ensures that standards are maintained so that everyone, including B&Q, can rely on timber being produced in ways which do not destroy forests or cheat the local people. B&Q used 283,000 cubic metres of timber in 1994 (8.7 per cent up on 1993), mostly from temperate forests as shown in Table 5.9.

The second project described in the video explains how environmental teams have been established in their stores thus involving their employees in the environmental work. This has many benefits because their people learn about environmental matters, they see opportunities for new initiatives, have a forum where ideas can be followed up and put into practice and where customers and the local community can benefit as well as the company.

B&Q have also carried out a supplier audit and set standards so they can grade their suppliers. Ultimately they need to achieve the set standards if they wish to remain an approved supplier. The way in which supplier grading is done is shown in Table 5.10.

Table 5.10 B&Q Summary of Supplier Grading

Grade	Description
A	Demonstrate environmental excellence, using a systematic, well-documented environmental programme and showing innovative responses to environmental issues.
B	Environmental policy and thorough review of issues has been provided. The policy is supported by an action plan with clear objectives and real progress can be demonstrated.
C	Key issues identified, a framework policy established which commits the supplier to achieving broad objectives and the intention to implement the policy with specific objectives is sincere.
D	Suppliers have returned a questionnaire but not yet provided a policy statement with a proper assessment of environmental issues.
E	Suppliers have returned a questionnaire but the responses indicate that B&Q is exposed to severe liability in one or more areas. Immediate action is needed to address the issues which do not conform to legal requirements and B&Q's environmental policies.
F	Failure to respond to the questionnaire, other approaches and offer no information.
N	New suppliers who have been sent a questionnaire which needs to be returned within 6 months.

The progress in persuading domestic, European and international suppliers across all products to adopt environmental policies is shown in Table 5.11.

Table 5.11 B&Q Suppliers with Environmental Policies

	Feb 93	Dec 93	Aug 94	Dec 94	May 95
Percentage	8%	33%	56%	91%	96%

Alan Knight, does not believe that any individual organisation or country can become truly sustainable alone. However, he does say:

We are determined to do as much as we can to ensure that we only buy from suppliers which are, like B&Q, committed to the global desire for all stakeholders to act responsibly. We will assist our customers to do the same. Mistakes will be made, but these will be outweighed by the benefits. The benefits for the environment are obvious and the commercial benefits, both short and long term, easily pay for the investment.

(ibid, p99)

RECLAMATION SERVICES LIMITED

This business has grown into one of Britain's largest architectural salvage companies with enormous stocks of building stone, Cotswold tiles, hardwood beams, fireplaces, floorboards, statues, roofing slates, gates and so on – all recovered from old buildings. They undertake controlled dismantling and demolition work. This includes buildings such as the railway station at Swindon, a Building Society in Gloucester, a hospital in London, and old manor houses. They undertake this work paying strict attention to health and safety regulations and in accordance with salvage business criteria which avoids dealing with stolen goods. They also work within a framework of ethical standards, environmental criteria which emphasises re-use and recovery of materials.

Their import and export business of building materials, garden statuary and architectural features is widely known and respected in the trade. They hold periodic auctions and have a stone shop for restoration work, employing five masons on site.

Reclamation Services is included as an example of a small business which has established a niche for itself as an environmental business, illustrating how new business opportunities are emerging for the enterprising entrepreneur.

HOW TO ASSESS STRATEGIC INTEGRATION OF ENVIRONMENTAL MATTERS

These examples coupled with those that are more briefly described in Appendix No 1 provide the basis for suggesting some criteria which indicate that a particular organisation has gone beyond having an environmental policy which runs parallel with their business strategy. The evidence that organisations can use to test their strategic integration is set out in Table 5.12. Organisations doing most of these things can have confidence that environmental management is an integral part of their philosophy, their strategic thinking and their operating practices. They will be demonstrating how ethical and environmental responsibility contributes to solutions which take us towards sustainable societies.

Table 5.12 Evidence of Strategic Environmental Responsibility

- A Board statement explaining how the strategic thrust of the business contributes to the creation of sustainable societies.
- A Board statement showing how the organisation fits into a sustainable society and how progress towards explicit goals will be measured.
- Acknowledgement that the health and safety of employees, customers and the community are the organisation's responsibility and take precedence over profits.
- Social, ecological and economic considerations have equal weight when devising product strategy, operating processes and buying in services.
- Purchasing policy avoids scarce resources, endangered species, habitat destruction, oppressive regimes and cheating suppliers and growers.
- Operating practices are geared to cut emissions, which are continually monitored, make efficient use of all resources and introduce more severe targets in each successive year. Toxic substances are given special care.
- Waste reduction is an integral part of the way the business is run and employees contribute to implementing 'more from less' and 'reduce, re-use and recycle.'
- Continuous improvement is an accepted goal across all functions.
- Packaging and transportation targets are set to reduce their adverse impact.
- Accounting practices increasingly include the external costs which are typically omitted or ignored.
- All employees receive appropriate environmental training and environmental performance is included in the criteria used for selecting, rewarding and promoting people at all levels.
- Social, environmental and financial auditing are given equal weight and cross impacts are made explicit.

CONCLUSION

This chapter has looked at the difference between having an environmental policy which is managed alongside business strategy and one that is integrated with it. Both can improve business performance, environmental health and quality of life. However, when social responsibility and environmental health are an integral part of business strategy a more healthy, responsible business emerges. Employees at all levels appreciate the opportunity to contribute to solutions which stop the damage and to offer ideas for better business. They are directly involved in environmental management as part of their work and demonstrate a collective commitment to sustainability. In this way organisations are capable of achieving the dramatic shift that sustainable development requires.

Businesses which adopt an ethical approach and enlightened self-interest can still innovate to bring about significant changes from which the community, the environment and the business all benefit. This may not be the only way to achieve sustainability but several organisations are demonstrating that it works and it could be more widely applied.

Figure X

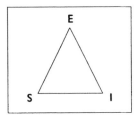

When this idea occurs in subsequent chapters attention will be drawn to the triangle model using the acronym ESI (pronounced easy!) E = Ethical, S = Enlightened Self-interest and I = Innovation. The shorthand symbol will be as shown above.

Some companies which are reaching out towards achieving a sustainable business have been described briefly. The progress being made by Procter & Gamble, Rank Xerox, Electrolux, The Co-operative Bank, B&Q and Reclamation Services is impressive. None of them has reached the goal of sustainable enterprise, as all would acknowledge, but the work they are doing is moving steadily towards that goal. All these businesses would probably agree with Alan Knight when he says that no single business can become sustainable on its own. That is why Chapter 6 considers how industry and professional associations are contributing and Chapter 7 looks at how communities are reaching out for sustainability.

6 INDUSTRY AND
PROFESSIONAL ASSOCIATIONS

There are several industry associations and many 'special interest groups' actively involved in developing the principles of sustainable development. Examples illustrate how professional bodies, industry associations, new groupings and emerging networks are making a difference to the way in which difficult issues are approached. These initiatives are collaborative ventures typically involving people with different interests. Some initiatives are major undertakings with worldwide implications, others are small scale. All seek to influence a significant group of people with special interests and reach a wider audience. The small scale examples emphasise the point that everyone can make a contribution.

CHEMICAL INDUSTRY

The chemical industry is perhaps one of the most environmentally controversial sectors. The process of making, distributing and using chemicals can be dangerous, posing risks to workers and nearby communities. Many of the industry's products can also damage the environment if misused or released accidentally.[1]

It is not surprising that the chemical industry was the first to take environmental matters seriously with their Responsible Care Programme. The Canadian Chemical Producers' Association (CCPA) was the first to develop the idea in the 1980s and adopted it in 1985. Despite slight variations the common features of Responsible Care Programmes are embodied in a statement which companies sign – see Table 6.1.[2]

Table 6.1 Responsible Care Programme

- A formal commitment to a set of guiding principles, signed by the Chief Executive Officer.
- A series of codes, guidance notes and checklists to help with implementation.
- Progressive development of indicators against which improvements in performance is measured.
- A process of communications with interested parties on health, safety and environmental matters.
- Provision of forums in which companies can share views and exchange experiences on implementation.
- Adoption of a title and logo that identifies products which measure up to agreed standards.
- Consider how to encourage all chemical companies to participate in the scheme.

Responsible Care was introduced as a voluntary initiative by the Chemical Manufacturers Association (CMA) in the USA and by the Chemical Industries Association (CIA) in the UK in 1988. Australia introduced a scheme in 1989, France in 1990 and Germany in 1991. Many other countries have followed with their own programmes including some developing countries and countries in the former Soviet Block.

The need for responsible care was strongly endorsed by Peter Sandman of the US Chemical Manufacturers Association, speaking in 1990, when he was quoted by Robert Kennedy Chief Executive Officer of Union Carbide in April 1991:

Your industry hasn't kept its emissions as low as you practicably could. Perhaps more important, you haven't built a good record of open communications with the communities in which you operate. Too often you have been arrogant or uncaring; sometimes you have been dishonest. [Robert Kennedy added: **'The result is that people don't trust us'.**[3]]

The Chemical Industry Association (CIA) in the UK has published its own booklet entitled *Chemicals in a Sustainable World.*[4] The booklet describes the elements of sustainable development and their view of the roles of government and industry. They go on to stress the need for cooperation between them and then consider the role of the community in achieving sustainable development. Legislation and economic instruments are discussed as are voluntary approaches. Practical ways of dealing with pollution, waste minimisation, product stewardship, energy conservation and transport are all described. Attention then focuses on management issues, international trade and sustainable lifestyles. The document concludes with a statement of chemical industry commitments and their recommendations as shown in Tables 6.2 and 6.3.

Table 6.2 Commitments by the Chemical Industry

In addition to compliance with the law, the chemical industry's contribution to sustainable development embraces Responsible Care which commits to:

- continuous improvement of industrial practices;
- measurement of performance and open communication of the results;
- development of measurable systems of environmental management;
- implement programmes of pollution prevention and waste minimisation;
- build and operate plants which protect employees, the public and the environment irrespective of national or international location;
- apply product stewardship and risk assessment to minimise the adverse impact of our products and processes;
- continuous improvements in the fields of resource conservation and fuel efficiency;
- select, monitor and manage safe and environmentally sound methods of transporting our products;
- support education systems, the sharing of experience and expertise, and properly controlled technology transfer as the bases of environmentally sound development; and
- be active in the debate on the selection and evaluation of legislation and other instruments which will be efficient and economically effective in promoting environmental behaviour change.

Most of the large chemical manufacturers have adopted change programme, are increasingly open with information and are actively striving to improve their reputation. The industry has formulated a set of recommendations which they would like all users of chemicals to adopt and these are set out in Table 6.3.

Table 6.3 Recommendations from the Chemical Industry

The chemical industry sees sustainable development as a partnership for which the following requirements are recommended:

1. the adoption by other industries of a Responsible Care approach;
2. the exploration of partnership agreements to bring about environmental improvement;
3. the development of long term national and international strategies which are independent of short-term political expediency;
4. recognition of the need for approaches that are scientifically, intellectually and economically sound;
5. provision of a climate for innovation and development of new technologies;
6. support for international legal and technical controls for protection of intellectual property;
7. commitment to the GATT route for the resolution of problems of international trade and the environment;
8. educational systems which incorporate throughout individuals' responsibilities towards health, safety and environmental issues; and
9. an evolutionary – not a revolutionary – approach to the application of legislative instruments.

The chemical industry wish to correct what they see as serious misconceptions about their industry and some of the information that they are now publicising comes from independent research which indicates that:[5]

- plants make their own chemicals to protect themselves against pests and this is why organically grown food often shows remarkable resilience. When this food is eaten the naturally produced chemicals are also consumed but the effects of eating these natural chemicals has not been tested;
- the incidence of breast cancer is not rising and the incidence of stomach cancer is falling;
- the use of chemicals has saved many lives and helped to increase food production; and
- environmental pollution is 30 times less significant as a cause of cancer than smoking, diet imbalance, chronic infections and hormones.

There is still a wide gap between those who are critical of the chemical industry and those within it who are seeking to improve its reputation, strengthen understanding of its positive contribution and improve the image of chemical companies. Paul Hawken suggests the need for major strategic changes when he says:[6]

If Du Pont, Monsanto and Dow believe they are in the synthetic chemical production business, and cannot change this belief, they and we are in trouble. If they are in business to serve people, to help solve problems, to use and employ the ingenuity of their workers to improve the lives of people around them by learning from nature that gives us life, we have a chance.

The criticism of the industry is taken much further by others and often amplified by the media when they seek 'scare' stories believing that they attract attention. Such stories include (italics are added to emphasise the caveats and uncertainties):[7]

- every year the developed world exports developing countries *many* millions of kilograms of *potentially lethal* chemicals which are too dangerous for use in the country of origin;
- the US National Research Council *estimates* that *up to* 20,000 Americans *may die* each year from *relatively* low levels of pesticides in *domestically* grown food;
- abandoned waste dumps, which *may include dioxins* can be chemical time-bombs as the chemicals work to the surface or contaminate underground water;
- *indiscriminate pesticide use* in agriculture has produced new strains of pest which are resistant to the commonly used sprays;
- chemicals such as DDT have been shown to concentrate in body fat. Small fish absorb tiny amounts, they are eaten by bigger fish which in turn are eaten by birds or people. At each stage in the food chain the ratio of DDT to body weight increases. *Sample rural workers in Central America* have 11 times as much DDT in their bodies as the average American. The governments in some countries continue to use these chemicals because *they believe the short term benefits outweigh the longer term damage*;
- there are over 70,000 chemicals in use and each year another 1,000 come into use. The US National Academy of Sciences conducted extensive tests on 100 randomly selected chemicals and *found from literature surveys* that only 10 per cent of pesticides, 2 per cent of cosmetic ingredients, 18 per cent of drugs and 5 per cent of food additives had *enough data* on them to enable complete health assessment to be made.[8]

The chemical industry is more open than ever with its information and more receptive to valid scientific evidence of the dangers than it has ever been. However, the public are more inclined to believe scientists employed by environmental groups than those employed by either companies or the government (see Table 3.3, p37). Over the years there have been many instances where the government, guided by business, has denied problems that have subsequently proved to be valid. Over the past 20 years this has occured with lead in petrol, asbestos, acid rain, threats to the ozone layer and BSE or mad cow disease. The UK government's 'culture of denial' was severly attacked by Geoffrey Lean, one of the UK's most respected environmental journalists.[9]

There are complex and difficult choices to be made. Sudden changes in the use of chemicals in farming can have devastating consequences as demonstrated by the beef industry after the scare over BSE and CJD. While the gulf in understanding between the chemical industry and environmental and health campaigners remain wide the whole community is at risk. There is an overwhelming case for much closer, scientific-based, dialogue to close the gap in understanding, resolve the concerns and find viable solutions which are good for the community the environment and the industry.

FOREST STEWARDSHIP COUNCIL

A combination of circumstances created the right conditions for the World Wide Fund for Nature to take an initiative to form the Forest Stewardship Council (FSC). The circumstances include absence of effective action on deforestation, a general lack of confidence that environmental claims are valid, and a reluctance to rely on legislation in a climate of deregulation. There was also an emerging consensus that business and environmental groups should work together on solutions and growing public concern about the state of the environment. The situation threatened businesses with demonstrations outside DIY stores and disquiet among shareholders that staff morale could decline and profits fall. The primary opportunity was to demonstrate that rain forests were not being destroyed and that the timber trade was behaving responsibly. The scene was set for a new initiative in the 1990s.

An initial difficulty arose over the definition of a sustainable forest. Those interested in the timber believed that 'sustainable yield' was enough but the FSC pointed out that there is much more at stake than quantity of timber – even if yields are sustainable.

The idea is to get members of the wood and wood pulp trade to take responsibility for their supply chains leading towards purchasing only from well-managed forests. In the UK the WWF 1995 Plus Group was formed to implement the FSC principles among firms dealing with timber in the UK. By March 1996 their member companies:[10]

- sold £2.4 billion worth of wood and wood products per year. This represents the business of 47 member companies;
- these companies represent 22 per cent of UK wood use;
- they bought 4 per cent of their timber (£97m worth) from independently certified forests;
- they have de-listed 99 suppliers who failed to provide adequate evidence of forest origin of their timber; and
- by 31 December 1999, 10 members will buy only wood products which have been approved through the FSC system.

Later the definition of well-managed forests was strengthened, a three star system was introduced, the requirements for membership were refined and six-monthly reports were requested. The FSC emerged as the primary source of good practice with targets set which were ahead of those set by the International Timber Trade Organisation (ITTO). The FSC now claims:

The Forest Stewardship Council (FSC) is the only independent, international body which can credibly guarantee that forest management is not damaging a forest, its wildlife and the people that depend on it.[11]

The commitments, targets and obligations of the WWF 1995 Plus Group are set out in Table 6.4. By July 1996 72 UK companies had joined and plans were being made to extend the principles worldwide.

Table 6.4 Commitments, Targets and Obligations of the 1995 Plus Group (agreed 5 July 1995)

1 Members are committed to the Forest Stewardship Council as the only currently credible independent certification and labelling system with global application.
2 Members are committed to the phasing-out of the purchase of wood and wood products which do not come from well-managed forests as verified by independent certifiers accredited by the Forest Stewardship Council.
3 In order to fulfil point 2 members will aim to be purchasing only certified wood products by 31 December 1999.
4 A named senior manager will have responsibility for implementing the above commitment and targets.
5 Progress towards the targets will be monitored via six-monthly progress reports. Reports must include a database of wood and wood products used by quantity, type and forest source. Sources must be categorised as (1) certified forest (2) known well-managed forest or (3) unknown and/or not well-managed forest.
6 Members of the WWF 1995 Plus Group may use the Forest Stewardship Council product mark when they are licensed to do so. Other labels describing the quality of management source forests will be phased out.

The FSC has been a success story. The logo came into use in early 1996 so that customers could choose to buy approved products. As yet the total quantity of approved timber represents a tiny proportion of the total timber used in the UK, but the achievements will become greater as the number of organisations involved increases. Success is attributed to:

- a defined and recognisable problem;
- companies support the scheme because their business is threatened;
- the FSC helps to turn the problem into an opportunity;
- the involvement of WWF adds credibility to the scheme (NB this is in contrast with the Chemical Industry approach which was an internal initiative); and
- collaborative schemes are increasingly recognised.

A complementary independent study on the sustainability of the pulp and paper industry is published by IIED and the World Business Council for Sustainable Development (WBCSD).[12] The report describes Sustainable Forest Management (SFM) which it emphasises is still evolving but includes sustainable yields of wood and non-wood products (for example nuts, fish and recreation services), protection of soil and water, maintenance of ecosystems, biodiversity, the health of the forest and positive benefits for local communities.

Best practices in paper manufacture are described and the different situation facing industrialised and developing countries. For example in developing countries paper consumption is 12 kg per person per year but in industrialised countries it is 152kg per person per year. The need for growth in the former is necessary while the latter could explore ways in which to bring down the levels of consumption. The recommendations cover action which could be taken to extend best practices more widely. Specific proposals are put forward for the paper industry, governments, international agencies, consumers, non-governmental organisations (NGOs) and investors.

The success of the FSC and SFM schemes is leading towards further improvements as more organisations get involved. The up to date Principles of Forest Management are shown in Table 6.5

Table 6.5 FSC Principles of Forest Management

1   Compliance with Laws and FSC Principles. Forest management shall respect all applicable laws of the country in which they occur and international treaties and agreements to which the country is a signatory, and comply with all FSC principles and criteria.

2   Tenure and Use Rights and Responsibilities. Long-term tenure and use rights to the land and forest resources shall be clearly defined, documented and legally established.

3   Indigenous Peoples Rights. The legal and customary rights of indigenous peoples to own, use and manage their lands, territories and resources shall be recognised and respected.

4   Community Relations and Workers' Rights. Forest management operations shall maintain or enhance the long-term social and economic well-being of forest workers and local communities.

5   Benefits from the Forest. Forest management operations shall encourage the efficient use of the forest's multiple products and services to ensure economic viability and a wide range of environmental and social benefits.

6   Environmental Impact. Forest management shall conserve biological diversity and its associated values, water resources, soils and unique and fragile ecosystems and landscapes, and, by so doing, maintain the ecological functions and integrity of the forest.

7   Management Plan. A management plan – appropriate to the scale and intensity of the operations – shall be written, implemented and kept up to date. The long-term objectives of management, and the means of achieving them, shall be clearly stated.

8   Monitoring and Assessment. Monitoring shall be conducted – appropriate to the scale and intensity of forest management – to assess the condition of the forest, yields of forest products, chain of custody, management activities and their social and environmental impacts.

9   Maintenance of Natural Forests. Primary forests, well-developed secondary forests and sites of major environmental, social or cultural significance shall be conserved. Such areas shall not be replaced by tree plantations or other land use.

10 Plantations. Plantations shall be planned and managed in accordance with Principles and Criteria 1–9, and Principle 10 and its Criteria. While plantations can provide an array of social and economic benefits, and can contribute to satisfying the world's needs for forest products, they should complement the management of, reduce pressure on, and promote the restoration and conservation of natural forests.

MARINE STEWARDSHIP COUNCIL

The demand for fish and seafood products has never been greater. In China consumption has doubled in the last decade and in the USA 3.94 billion pounds (1.79 billion kg) of fish were consumed in 1994. In the developing countries nearly 200 million people depend on fisheries for their livelihood.[13]

In contrast the world fish catch remained steady from 1988 to 1993, but rose in 1994 with large catches in China, Peru and Chile. Fish stocks are disappearing and fish are being caught at levels that are greater than can be sustained. Fishing fleets go further out to sea for their catches, in larger boats which stay at sea for months. Countries like Spain with one of the largest fishing fleets have had disputes with Canada and the UK. The larger fishing fleets also encroach into the coastal waters of developing countries and devastate the local fishing grounds on which their people depend. The total world fish harvest and the proportion that is from fish farming (aquaculture) is shown in Figure 6.1.

Figure 6.1 World Fish Harvest[14]

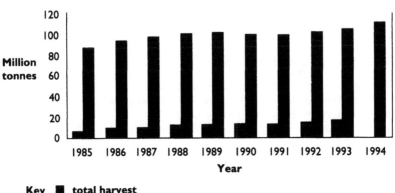

Key ■ total harvest
■ aquaculture

With world population rising the annual average world fish harvest per person remained fairly steady from 1964 until 1983, then rose to 19.4 kg per capita by 1988. Since then it has ranged between 18.1 and 19.3 kg per person. The Food and Agriculture Organisation (FAO) whose figures are quoted in the WWF Endangered Seas Campaign suggest that with aquaculture the total world fish catch could increase to 120 million tonnes per year by the year 2015. This assumes that grain supplies for feeding fish in fish farms will be available. This may be optimistic when the world grain harvest struggles to keep pace with demand. The problem is compounded by erosion of top soil, fertiliser usage is no longer increasing, water shortages are becoming more severe and climate disturbance cause crop failures on ever larger scales.[15] However, it requires only 2 kg of grain, or less, to produce 1 kg of live weight gain in fish, compared with 2.2 kg for chicken, 4 kg for pork and 7 kg for beef.

The endangered seas project and the formation of the Marine Stewardship Council (MSC) is timely. The MSC has been formed by WWF and Unilever, the Anglo-Dutch consumer goods group. Unilever is a massive worldwide business with a £25 billion turnover of which fish and fish products represent £600 million per year. Unilever has pledged that all its fish products will be accredited by the MSC by the year 2005.[16]

In addition to Unilever's pledge the MSC's challenging task is to persuade other major organisations involved in the fishing industry to join the MSC and make similar pledges. This is proving successful with forestry but the task with fishing is, if anything, even more daunting. In many countries the fishing industry is heavily subsidised, enormously destructive, the fleets exceed requirements for an industry that needs to conserve stocks and it is a very large employer. The importance of fishing to coastal people is evident throughout the developed and developing world.

Michael Sutton of WWF and Caroline Whitfield of Unilever in a joint statement say:[17]

To reverse the fisheries crisis, we must develop long-term solutions that are environmentally necessary and then, through economic incentives, make them politically feasible... In order to make this work, the conservation community and progressive members of the seafood industry must forge a strategic alliance... The MSC has the potential to significantly alter worldwide fishing practices in favour of more sustainable, less destructive fisheries.

(p3)

A summary of the objectives of the MSC are shown in Table 6.6.

Table 6.6 Summary of the MSC Objectives

- To develop the MSC partnership as a non-profit non-governmental organisation.
- To ensure the long-term viability of global fish populations and the health of the marine ecosystems on which they depend.
- To establish a broad set of principles for sustainable fishing and set standards for individual fisheries.
- To establish an independent accreditation scheme.
- To have products which are accredited marked with an on-pack logo by 1998.
- To agree the principles and standards through regional consultations.

INTERNATIONAL HOTELS ENVIRONMENT INITIATIVE

The International Hotels Environment Initiative (IHEI) was formed in 1992 with John van Praag of Intercontinental Hotels playing a significant role in its formation and obtaining formal endorsement from Prince Charles. Its objectives are set out in Table 6.7.

Table 6.7 Global Aims of IHEI

1 Facilitating access for hotels to environmental information and techniques.
2 Highlighting and networking best practice from international hotels.
3 Activating champions and 'partners for action' throughout the world.

The scope for hotels to take action which enhances their business and at the same time improves the environment is enormous. Every hotel visitor on arrival at a hotel forms an immediate impression. The grounds, booking-in procedure, lighting and heating, bedrooms and the calibre of the staff are all noticed even if only subconsciously. Poor hotels are easily recognised not only from lack of service but also cleanliness, quality of food, comfort and convenience. With environmental concern rising up the agenda, increasing numbers of visitors notice how one hotel differs from another. This can include such things as efficient heating and lighting, avoidance of excessive use of chemicals in the grounds, the option to not have bed linen and towels changed daily and the awareness of staff about environmental matters.

These are typical issues which are now emerging at the forefront of best practice in the hotel industry. By taking care of the hotel environment hotel groups can improve their competitiveness. Strategically, hotels can also consider new business opportunities which may arise as a result of better environmental management, such as:

- Does the location help save customer transport costs? An overnight stay may be cheaper and cause less environmental damage than going home and returning the next day?
- Are hotel managers trained to appreciate environmental factors, and are these included in their performance targets?
- Are staff aware of and committed to making their contribution to both business results *and* environmental performance?
- Can the hotel legitimately use its environmental performance for publicity?

In order to help hotels develop and implement their own environmental programme and to publicise best practice the IHEI makes available several aids for the hotel and hospitality industries. These include a user-friendly action pack, an environmental management video, a standard reference to best practice, a multilingual directory of practical environmental information, and the quarterly magazine of the IHEI.[18]

ENGINEERING PROFESSION

In June 1995 The Royal Academy of Engineering was one of 14 national engineering organisations that formed the Council of Academies of Engineering and Technological Sciences (CAETS, founded in 1978) which published a joint Declaration on *The Role of Technology in Environmentally Sustainable Development*.[19] The Declaration highlights the important role of technology in a sustainable future for humanity. Their beliefs are given in Table 6.8.

The Declaration states the conviction that sound economic development is possible only if full use is made of environmentally advantageous technologies. The transfer of technology to developing countries is recognised as an important factor. It highlights in particular the following areas where technology has a crucially important part to play:

- Energy: the need to reduce the use of fossil fuels, to improve the efficiency of automobiles, to develop renewable energy sources;

Table 6.8 CAETS: Beliefs Contained in their Declaration

- Engineering and technology are key components of national and international efforts to achieve environmentally sustainable economic development for all nations.
- The goals of economic development and environmental protection can be compatible.
- The ability to achieve global sustainability is strongly influenced by population growth.
- The variability of natural systems is easily confused with the impact of human activity on the environment. This is an important subject for research.
- For environmental issues such as those involving the oceans and atmosphere, the planet must be regarded as a single complex system whose interests are best served by international action, involving developed and developing nations.
- Most actions to achieve environmentally sustainable economic development will take place at the national level, while taking into account the interdependence of nations.
- The achievement of environmentally sustainable economic development depends on adequate technical education at all levels and balancing economic, social and environmental goals.

- Public infrastructures: which include water resources, power systems, bridges, roads, communications and transport facilities;
- Water: the treatment and re-use of water, which will have a decisive role in sustainable development including public health, agriculture, industry, controlling pollution, minimising water consumption and avoiding waste;
- Food: increasing food production, improving storage and distribution. The development of biotechnology, reduction in use of toxic chemicals and sustainable farm practices;
- Manufacturing and mining: industry has begun to reduce, re-use and recycle materials, the search is on for industrial ecosystems that imitate natural ones, product and process design is taking more account of environmental factors, renewable agriculture and forestry is expanding and degraded landscapes are being rehabilitated;
- Materials: steel, concrete and plastic are changing to reduce their environmental impact, new materials are being designed which are more energy efficient, consume less mineral resources, are lighter and stronger and superior to other materials; and
- Information technology: information technology has the potential to alter how and where people live, change the nature of urban areas, modify the way in which organisations are managed and improve the efficiency of air, land and water transport. Fibre optic cables and Earth-orbiting satellites are extending our ability to survey and protect the environment, minimise pollution and improve energy efficiency.

The report acknowledges that the choice of appropriate solutions will depend on governments, international agencies, consumers, private industry and educational institutes as they struggle to balance the needs of the environment with social acceptability and economic considerations. It concludes with a set of recommendations for action which are addressed to governments, industry, international agencies, educational institutes and engineering institutions.

In September 1995, just three months after the CAETS Declaration, The Royal Academy of Engineering held their Conference *Engineering for Sustainable Development*.[20] The Chairman, Sir William Barlow, confirmed in his closing remarks that sustainable development is a core theme in the Royal Academy's Corporate Plan and the conference marked the start of a process 'to galvanise engineers into action.' The President of The Royal Academy reiterated Dame Rachel Waterhouse, former Chairman of the Consumers' Association, when she said 'Engineers must take centre stage, don't be afraid'. The summary statement about the conference picked up several themes which recurred during the conference to give them special emphasis. These are noted in Table 6.9.

Table 6.9 Engineering for Sustainable Development – Key Themes

---

- Engineers have a contribution to make to improve the quality of life for all through technological solutions and can be prime movers in sustainable development.
- Engineers are well placed to disseminate best practices and accelerate the pace of sustainable development, especially to developing countries.
- Market opportunities are being created as the demand for technological solutions increases.
- Clean technology and sustainable development are not the same thing, the latter being more wide-ranging and including, for example, rate of consumption.
- Engineering design needs to move from fitness for purpose based on engineering rationality to recognise social priorities and sustainable considerations.

SOME OTHER IMPORTANT INITIATIVES

In addition to the major initiatives described above there are many other examples of innovative approaches – large scale and small scale – which contribute to the creation of sustainable communities. The examples below describe briefly some of these initiatives because even a brief reference helps to emphasise the diversity of contributions to the principles of sustainable development.

Building and Construction

The Building Research Establishment (BRE),[21] mainly funded by the Department of the Environment, is the UK's leading source of independent information and advice on the safety and efficiency of buildings, how they can be improved and on their construction. The BRE contributes to building standards and codes in Europe and internationally.

BRE has established The Building Research Establishment Environmental Assessment Method (BREEAM) which provides independent assessments for particular buildings.[22] This enables developers and other potential clients to communicate the environmental performance of a building in a credible way. The assessment covers such things as carbon dioxide emissions, air quality, re-use and recycling of materials, on-site ecology, noise and lighting.

BRE in conjunction with the PA Consulting Group have developed *The Office Toolkit** which is designed for office and facilities managers to reduce the costs of managing buildings in ways that help the environment. As a result of several successful schemes it is estimated that it is possible to save up to £200 per employee by managing buildings and facilities more efficiently and more responsibly.

The advice emanating from BRE for new buildings is being applied on their own doorstep. Their new office and seminar block under construction at Garston will use recycled aggregate – the first time this has been done in the UK. This is a particularly significant development and a key feature of sustainable construction. The new BRE building is targeted to use 30 per cent less energy than current good practice. It will provide comfort and high levels of environmental control with natural cross-ventilation and a novel floor, incorporating water cooling.

BRE want to get others involved in improved building design. Partners in Technology is an example of a BRE initiative with 18 industry partners. The aim of this three-year project is to identify the effects of building materials on the environment over their full life cycle. The partners come from a wide cross-section of manufacturers involved in construction materials including aggregates, clay pipes, PVC, timber, cement and metals.[23]

Electricity, Electronics and Communications

The number and the size of worldwide companies involved with electricity is astonishing. The Times 1000 lists the World's Top 50 Industrial Companies and the summary by industry sectors is shown in Table 6.10.[24] There are four sectors, namely electricity, communications, electronics, and electricals which are directly associated with electrical power generation, its use or manufacturing and marketing equipment which depends on it. These four sectors (23 companies) account for nearly half of the capital employed by the world's 50 largest corporations.

Table 6.10 World's Top Companies by Sector

Industry	Number of companies	Total capital employed £1000
Transport	9	308,244,107
Electricity	9	267,298,771
Oil, gas and nuclear fuels	9	230,386,369
Communications	8	220,177,236
Electronics	4	104,265,685
Electricals	2	89,275,976
Other	9	164,209,040

* Brochure available from CRC Ltd, 151 Rosebery Avenue, London EC1R 4QX; Telephone 0171 505 6622.

The four 'electrica' sectors are changing the face of industry worldwide by the remarkable rate of technological progress and the rate of growth of these industries. An example of the speed of growth was quoted by Chris Tuppen, of British Telecommunications when he said: 'In 1915 it took 18 kilograms of copper per kilometre of cable to make a telephone call. Today it takes 0.001 grammes of glass per kilometre of cable to make the same telephone call.'[25] This point was made in the discussion following three papers on The Challenge for Manufacturing at The Royal Academy of Engineering Conference in September 1995. Dr Peter White of Procter & Gamble referred to their slogan *more from less* (mentioned in Chapter 5) and went on to explain that BT's example demonstrated the validity of the slogan. He then described how technology can bring about a paradigm shift – in the BT case when fibre optics creates an alternative to copper.

It appears that the electronics industry has not published a policy statement on sustainable development.[26] One major reason for this is that the rate of change is so rapid. The growth rate is believed to be between 5 per cent and 20 per cent per annum for different parts of the industry. With this rate of change any statement can become dated very quickly. However, the industry is consciously encouraging progress towards sustainable development in several ways such as the development and use of solar cells and the use of mobile telephones in remote areas which saves mining more copper for telephone lines.

Major achievements are being made in dramatically reducing the quantity of energy required for each unit of manufacture – across a range of industries. There is also great scope for leading the way with renewable sources of power, especially solar and wind, but also helping to make use of localised potential for tidal and geothermal energy. Despite the rate of change there is scope for a regular summary statement of achievements and sharing of ideas under development across the whole industry.

Institute of Grocery Distribution (IGD)

An impressive example of a partnership initiative is the work done on recycling in Adur District and Worthing Borough on the South coast of England. This shows how innovation can deliver cost-effective sustainability improvements. The IGD Integrated Waste Management Working Group is led by Peter Hindle, who is also Procter & Gamble's European Director of Environmental Quality. P&G became involved with this scheme because they were aware that continuing improvements in environmental management needed to look at the disposal of their products' packaging after use. However, this is not something they could deal with alone.

The UK government set a target in 1990 that by the year 2000 25 per cent of household dust bin contents should be recycled. A pilot scheme was set up in Adur supported by the European Recovery and Recycling Association (ERRA), of which P&G are a member.

Householders were provided with boxes for their recyclable waste which were filled and put out for collection. The scheme was welcomed by householders and over 80 per cent took part. Those for whom the box scheme was inappropriate could use mini-recycling centres. Only when composting facilities were added to packaging and newspapers recycling did Adur reach the target of 25 per cent of household rubbish. After five years this level of recycling has been kept up so it is not a passing fad.

The additional cost of the waste collection service in Adur was £17 per household per year when compared to disposal to landfill. If the whole country adopted this scheme it would cost almost £400 million per year. Since the start the cost of the service in Adur has been brought down to £10 per household per year by introducing more cost-effective methods.

In 1993 the IGD was invited by ERRA to use what had been learned in Adur to devise a scheme for Worthing with a target cost of less than £5 per household per year The Adur experience resulted in practical learning in many areas. This included the right size of box to use, how to persuade householders to take part, the value of the kerbside box as a visual reminder about the scheme, the value of mini-recycling centres for people living in flats, the importance of including newspapers and composting and how to sort tin cans from aluminium cans and plastic bottles. In Worthing the collection system had to be redesigned so that all household rubbish – recyclable and non-recyclable – could be collected by the same vehicle, instead of separate vehicles as had been done in Adur.

The collection arrangements and the design of a suitable vehicle with several compartments involved other organisations and widened the partnership. The facility where sorting of bottles from cans takes place has been refined and can sort 250 bottles per minute into different plastic types and colours.

The costs in Worthing are remarkable. The additional cost per household per year is down to £1.33 per year – or £30 million per year for the whole country. Further cost savings are believed to be possible. As Peter Hindle says: 'For less than the cost of one lottery ticket per household per year we can deliver the household waste recycling that people are demanding'.[27]

Worthing is the test market that has proved that the learning from Adur can be applied elsewhere in a cost-effective and sustainable way. The challenge now is to extend the scheme, initially to the whole of West Sussex and then to the country and Europe.

Insurance

The Insurance industry has faced heavy losses in recent years as a result of disasters caused by extreme weather conditions. In 1995 the industry published their own statement of environmental commitment supported by the United Nations Environment Programme. It was a popular move and 50 major insurance companies had endorsed the statement by early 1996.[28]

The Rocky Mountain Institute's (RMI) Newsletter picked up this theme pointing out that global warming and climate change do not have to be a certainty to be a risk. It is interesting that the insurance industry is becoming the single biggest lever for action on climate change. They can influence governments'schemes for using energy more efficiently and stimulate the development of alternative technologies.[29]

Figure X

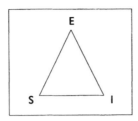

Ralph Cavanagh of the US Natural Resources Defense Council, says 'We're trying to align economic self-interest with solving the greenhouse problem.' RMI believe that companies will move quickly as soon as they 'see the financial rewards of abating the greenhouse problem.'

The Insurance industry worldwide receives £92,000 million ($1.4 trillion) each year in premiums. If a small proportion of this was invested in non-polluting energy alternatives, thereby accelerating the development of renewable sources, it could have a remarkable effect by making, for example, solar energy much more competitive with fossil fuels.[30]

Organisation Development, Learning and Change

Small scale initiatives can also make a contribution. The Association for Management Education and Development (AMED) has some 1600 members from public and private sector organisations, academia and consultancy. AMED is a unique association of individuals committed to developing people and organisations. AMED explores new ideas, shares research and good practice and develops skills needed for the future, for the personal and professional benefit of people in all communities. AMED's values and style include a strong emphasis on the management of individual and organisational learning and change. Members make a significant contribution at all levels by helping people to understand and cope with the impact of change on themselves and their organisations.

Some AMED members founded The Green Network in 1991 and this became the Sustainable Development Network (SDN) in 1994. Its small but active membership is open to non-members as well members of AMED. The SDN helps its members contribute to the development of sustainable societies through individual and organisational change. It publishes a newsletter every three months, provides its members with information on all members' interests and experience, what they want from and can offer to the network, and encourages networking. Periodic publications such as *Our Changing World: An Introduction to Sustainable Development* are produced and a variety of events are run in different parts of the UK.

Another initiative focuses on research. Professor Richard Welford has been a leading figure behind the formation of the International Sustainable Development Research Network (ISDRN), founded in 1995 with some 350 members by summer 1996.[31] Welford is also involved with five environmental management journals which are published by John Wiley & Sons.[32]

CONCLUSION

A surprising amount of activity on environmental matters and sustainable development has been initiated in the mid 1990s. This is a most encouraging sign from many sectors of society. It shows a general recognition not only of the importance of this subject for humanity but also articulates how different professional or industry associations can make a contribution. These initiatives are not confined to the UK, which is at the leading edge on some matters and trailing woefully on others.

The statements from industry associations and professional groups quoted here endorse the importance of sustainable development. It is increasingly clear that this requires attention to economic viability, social responsibility and ecological sanity. It is not just an environmental issue and the environmental problems cannot be solved in isolation. However, environmentally driven solutions are also producing positive results through improved efficiency.

It is striking to note that enlightened self-interest and an ethical approach are widely quoted, directly or indirectly, as the twin foundations for progress. Innovation is then needed to produce new solutions. Sometimes, as in the case of electronics, the rate of innovation is staggering. This demonstrates how crucial some forms of technology are for future, sustainable progress. Fortunately many enthusiastic technologists also recognise the significance of the social and ecological dimensions.

The emphasis in this chapter has been on those associations which have made clear progressive statements about their role and contribution. Other important areas have not been covered, such as transport, tourism and water, all of which are crucially important. Space is restricted, but sometimes the reason for omission is that it has proved difficult to obtain an appropriate policy statement or declaration of intent. This may be because it has not been written, that insufficient research has been carried out or that there is no champion to move the issue forward. The hope remains that the examples quoted will help stimulate action in new areas, led by innovative and enlightened champions, from which we can all benefit.

7 INDUSTRIAL ECOLOGY AND
COMMUNITY RENEWAL

In addition to the initiatives being taken by individual organisations (Chapter 5) and by industry associations and professional groups (Chapter 6) a variety of initiatives are being taken under a broad heading of 'industrial ecology' and within communities. In this chapter industrial ecology is considered first, followed by various kinds of community renewal and then some interesting examples of visioning the future. These examples are brought together in one chapter because they share a common characteristic of cooperation between people, businesses, voluntary organisations and local authorities. They are designed to be economically viable, to enhance environmental health, reduce environmental damage, provide employment, be socially responsible and make progress towards sustainable communities. Examples are used to indicate the creative ways in which people in different communities are making progress towards new, innovative solutions. The starting points and the action taken vary, showing that there is no single best way forward.

INDUSTRIAL ECOLOGY

The best known example of industrial ecology is probably Kalundborg in Denmark which uses the term 'industrial symbiosis' to describe the initiative which began in the 1970s. Businesses in Kalundborg have devised a system for exchanging resources so that wastes from one business are resources for another. In this section we then go on to look at other variations in Austria, The Netherlands and North America. It should be noted that some industries, notably chemicals, have for many years used the wastes, or by-products, of one process as raw materials for another, but these are often contained within a single business. The Industrial Ecology examples considered here involve cooperation between several different industries. Table 7.1 provides a brief description of industrial ecology.

Table 7.1 Industrial Ecology

Industrial ecology involves the application of natural ecological processes to industry. In the natural world, which includes human biological processes, there is no waste because everything is absorbed somewhere. Nutrients required by one species are obtained from other species of plants, animals and insects as they die and decay.

Efforts to apply this principle to industry is called industrial ecology. When it works well wastes are re-used, energy use is efficient, toxicity is minimised, damage to the natural environment is reduced, businesses benefit from cost-effective procurement of resources and avoidance of costly disposal to landfill or incineration and the community benefits from a healthier environment.

Kalundborg, Denmark[1]

In Kalundborg each resource exchange is a separately negotiated deal which makes economic sense to both parties. The solutions they have devised are from enlightened self-interest but they also improve the environment and benefit the local community. Devising the solutions required a willingness to innovate and although not motivated by ethical considerations they are ethically sound. It is worth looking at Kalundborg more closely.

Figure X

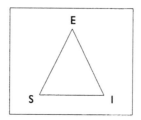

Major achievements often have small beginnings and Kalundborg, situated some 70 miles west of Copenhagen, demonstrates this principle. Asnaes Power Station, built on reclaimed land is Denmark's largest power station, providing electricity to the Zealand grid. Asnaes Power Station has a capacity of 1500 megawatts which means that it can supply 60 per cent of the demand for electricity in Eastern Denmark.

Electricity generation requires large quantities of water which came from natural underground sources. In the early 1980s the power station started supplying process steam to the Statoil refinery and Novo Nordisk both of which are located nearby. The Statoil Refinery is Denmark's largest with a capacity of 5 million tons of crude oil a year and a strong desire to be a leader in sound environmental management. Novo Nordisk is one of the world's leading biotechnology companies and a major producer of enzyme products with plants in some nine countries. Novo Nordisk aims to replace chemicals used in agriculture, pulp and paper manufacture with their own less damaging products.

It soon became apparent that the Asnaes Power Station was depleting ground water faster than natural processes could replenish it so, Tisso, a water utility became the source of supply of surface water, not only to the power station but also to the oil refinery and the enzyme manufacturer. This greatly reduced the use of ground water. A further reduction in the use of ground water occurred when the oil refinery, which used to discharge its waste water into the Kalundborg Fjord, started to supply it to the power station for use as cooling water. This now exceeds 1.5 million tons per year.

Since the 1980s Asnaes Power Station has provided district heat to Kalundborg (population around 50,000) and this has led to replacing 3500 oil-burning domestic heating appliances. Waste heat, previously discharged into the fjord, is used in a fish farm which produces 200 tons of trout a year for the French market. The fish grow more rapidly in the warmer water. Sludge (330,000 tons per year) from the fish farm's water treatment plant is used as fertiliser by farmers.

When the power station was built it was coal fired but to add flexibility it has been adapted to burn gas as well and can now use cleaned gas that used to be flared by the oil refinery. The oil refinery has been able to reduce the flaring so that it is little more than a pilot light. This has reduced the power station's coal consumption by 30,000 tons a year.

The flue gases from the power station are scrubbed and a filter removes 99 per cent of the ash and dust particles. Fifty tons of ash per hour is produced and is used by Alborg Portland in the manufacture of cement and concrete and for road building. Nitrogen oxide emissions have been reduced by 50 per cent when compared with the original emissions from burning coal. Desulphurisation removes more than 95 per cent of the sulphur dioxide in the flue gases. This is washed with lime-water to produce gypsum – 250 tons every 24 hours. The gypsum is sold to and used by the Gyproc Group, the largest manufacturer of plasterboard in Denmark. Also, 3000 tons of pure sulphur are sold by the refinery to a sulphuric acid manufacturer, Kemira. A project still under consideration is to use more of the waste heat in a massive greenhouse.

The local environment council has representatives from the businesses involved, farmers and people from the municipal authority. The local community appreciate the way in which this cooperation and exchange of resources has resulted in an improved environment with lower emissions and less demand on 'new' as opposed to recycled resources. One of the unique characteristics of Kalundborg is that there is widespread awareness of what is being done and how it helps the community and the environment. The achievements at Kalundborg are an excellent example of ESI in action. See Figure 7.1.

Information about Kalundborg has gone around the world and an extensive bibliography is available.[2] Jorgen Christensen, until recently Vice President of Novo Nordisk, Kalundborg, who was one the influential people in the industrial symbiosis, has identified several conditions which made a significant contribution to the success of Kalundborg, and may be important elsewhere:

- the industries should be different but fit together;
- negotiated deals must be commercially sound for both parties;
- collaboration with local authorities is necessary;
- partners should not be too distant because it adds costs;
- it helps when the managers know each other;
- appealing ideas outside the 'core business' should be resisted; and
- reliability and risk minimisation are extremely important.

Learning from this successful innovation has been remarkably slow to spread. This could be due to lack of appropriate support from government, the general reluctance of different businesses to exchange information and the absence of pressure from local communities.

Figure 7.1 Industrial Symbiosis (NB Original is colour coded)

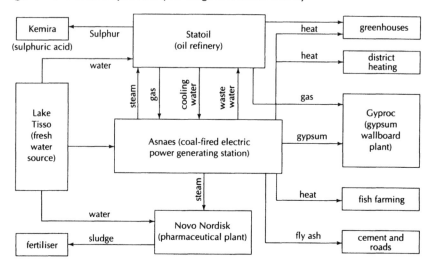

Source: *Industrial symbiosis*, brochure by Asnaes Power Station, Gyproc, Novo Nordisk and Statoil, Kalundborg.

Styria, Austria

For many years no other examples were known despite empirical studies carried out by Erich Schwarz and others in Austria and Germany.[3] However a network of firms exchanging resources does exist in Styria, Austria. It is much larger than Kalundborg but there is no overall intergrating concept comparable to Kalundborg. As Schwarz says 'no recycling network could be identified in which the participating companies pursue a common superior goal.'

The Styria network of 'recycling cells and structures' involves over 50 organisations which include agriculture, food processing, plastics, fabrics, energy, paper, wood working, building materials, metal processing and a variety of waste management organisations.[4] The difference with Kalundborg is that each exchange is a discrete transaction which makes current economic sense. This means that there is no possibility of further development other than more economically viable, independent transactions.

Eco-Industrial Parks in the USA[5]

Eco-Industrial Parks (EIP) are spreading in the USA. The goals of EIPs are to seek 'enhanced environmental and economic performance' by working together. This is achieved by applying the principles of industrial ecology (see Table 7.1). With sound design it is possible to devise ways in which raw materials can be saved, energy used more efficiently, water consumption reduced and support services made more efficient. This can be done more easily when establishing new developments but retrospective adjustments are also possible. Indigo Development, a consultancy based in Oakland California, encourages the use of the term EIP when there is:

- more than a single resource exchange;
- a network of different industries and services are involved;
- more than a single environmental theme, for example a solar energy park;
- sound interaction between businesses and the natural environment; and
- more efficient use of energy, materials and water.

When establishing an EIP it is desirable to keep an open mind about the best way forward. By involving the interested parties it is possible to explore design options for exchanging resources in cost-effective ways, examine where there are opportunities to 'close the loop', and identify opportunities to share support services. One of the aspects of an effective park is the willingness to share information and cooperate with different businesses. This is best done by the executives within the businesses but it helps to have some form of park management to provide facilities, events and support processes. Creating awareness of the types of industry and service organisations in a park is important and it helps if a notice board or more active method of communication is used to enable people to identify others who wish to explore possibilities.

Eco-industrial Parks have been established in Nova Scotia, Canada; Chattanooga, Tennessee; Brownsville, Texas, Baltimore and The Port of Cape Charles. Maybe there is scope to develop something similar in other countries but local circumstances and cultural considerations would need to inform the way in which this done.

Lessons from Industrial Ecology[6]

The lessons from industrial symbiosis in Kalundborg, the recycling network in Styria, the eco-industrial parks in Canada and the USA and industrial ecology in general can be summarised as shown in Table 7.2. However one general principle should be noted first. Before the advent of the interest in industrial ecology many businesses

Table 7.2 Lessons from Industrial Ecology

- If pollution can be stopped at source that is the best solution.
- The broad aim of industrial ecology is to mimic nature so that wastes become useful resources for another process.
- Good design is important and best considered at an early stage.
- The use of virgin materials can be reduced significantly.
- Pollution and wastes sent to landfill or incineration are cut to a minimum, thereby saving money.
- Energy and water are used efficiently with resulting cost savings.
- Retrospective modifications as well as new design is possible.
- More use is made of renewable energy sources.
- Negotiated deals which make economic sense for both parties improve the environment with benefits for the community.
- Innovation can result from focusing on product development as well as optimising the flows of energy and materials.
- A master plan may be useful but is not essential at the start (there was no master plan at Kalundborg or Styria).

Adapted from Ernie Lowe, Indigo Development, Oakland, California discovered for themselves that stopping pollution at source is invariably better for the environment and usually cheaper for the business especially when the risks of accidents and the escalating cost of waste is taken into account. One danger of industrial ecology is that attention will focus too readily on finding another way to use the wastes and emissions when the better option is to avoid them in the first place.

The principles of The Natural Step (Chapter 4), industrial ecology, waste minimisation, ethics and enlightened self-interest could usefully converge. Additional stimuli could come from government regulatory and fiscal measures which make waste disposal increasingly expensive. These two sets of influence could encourage the appropriate development of industrial ecology in many more countries, cities and towns.

COMMUNITY RENEWAL

In this section we look at examples from Brazil, the application of Local Agenda 21 in the UK and economic renewal in communities in the USA.

Curitiba, Brazil

'We live in the best city in the world.' According to Donella Meadows this is the belief of the residents of Curitiba, Brazil.[7] Small beginnings can lead to great achievements. In the 1950s Curitiba had a growing population of 150,000. In the 1960s the mayor sponsored a competition to find a master plan for the town and among those who responded was a group of young architects. They were critical of borrowed money being spent on grandiose schemes such as big roads, massive buildings and shopping malls.

Jaime Lerner was one of those architects and in the early 1970s he was appointed mayor by the military junta ruling Brazil at the time. By this time the population of Curitiba had risen to 500,000 and the problems were getting more severe with the spread of slums (*favelas*), accumulation of garbage, periodic floods, and poor transport systems. Because of Brazil's economic situation Lerner knew that any viable solution had to be cheap, small scale, relevant and should involve people in establishing the solutions.

Figure X

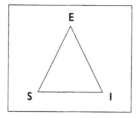

The solutions included planting 1.5 million tree seedlings, diverting water into new park lakes from low lying areas and employing teenagers to keep the parks clean. A 30-day trial for a shopping precinct in the downtown area proved so popular that other streets wanted to be included and many of these are now lined with gardens which are looked after by the children. This is a good example of ethically appropriate solutions (E) which matched the self-interest of the community (S) and achieved results through innovation (I). Creative solutions for Curitiba do not stop there – many practical solutions have been devised giving several more examples which can be interpreted using the ESI model.

All over Brazil there is a major problem with abandoned or orphaned children. Lerner arranged for industries, shops and institutions to adopt a few children and provide them with a meal in exchange for small tasks – but the dangers of exploiting child labour need to be avoided. Another innovation was to organise street vendors into open-air, mobile fairs which circulated in the city.

There was still the problem of population growth and Curitiba now has over 1.6 million inhabitants with increasing quantities of garbage. City centre streets are too narrow for trucks to get in to collect the rubbish, so another innovative solution was required. All Curitiba citizens separate their rubbish into organic and inorganic. Poor families bring their rubbish to 54 neighbourhood centres where they can exchange it for food or bus tickets. This helps to improve the living standards of the poor and helps to prevent disease. It has also cut the cost of waste collection and 70 per cent of rubbish is now recycled.[8] The rubbish is taken to the recycling plant where further separation of glass from cans is done by handicapped people, recent immigrants and alcoholics, thereby helping some of the more disadvantaged people in the community at the same time as improving the environment. This could have widespread application in many emerging economies.

Another major achievement in Curitiba is an efficient public transport system. Circular bus routes connect with radial routes. On these radial routes the buses have three coaches, travel in their own traffic lanes and carry 300 passengers at a time. They are as fast as underground trains but at one-eightieth of the cost of construction. The buses stop at plexiglass stations, designed by Lerner, where passengers pay their fares before boarding which means that loading and unloading is quicker. Bus fares are low, waiting is kept to a minimum so this mode of travel is preferred by 70 per cent of commuters and shoppers.

It is not surprising that Lerner, originally a mayor appointed by a military junta has been elected mayor since democratic elections were restored in Brazil in 1984. The real achievement of Curitiba is that people feel proud of their city and responsible for it. As Lerner says:

'When we provide good buses, schools and health clinics, everyone feels respected... the first priorities are on the child and the environment.'

(Donella Meadows, 'A City of the Future', *Resurgence* 168, 1995)

Lerner sees no reason why similar results cannot be achieved elsewhere – it is a matter of 'going against the flow and making a human city'. (*The Guardian*, June 5 1996).

Economic Renewal in the USA

The Rocky Mountain Institute (RMI) in the USA has been advocating economic renewal to encourage communities to apply the principles of sustainable development. RMI are better known for their work on renewable energy, cutting down the dependence on fossil fuels (especially oil and gas) as a contribution to national security and their invention of the Hypercar. (See Chapter 4, p61).

RMI recommend that communities can be improved by undertaking a well organised economic renewal programme.[9] This means stimulating the desire to take action within the community and identifying practical projects which will make a difference. Most economic development initiatives are large scale, devised by 'experts' outside the local community and not assessed in terms of sustainability. The strongly encourage mobilising people in the local community to take responsibility for economic renewal focusing on the benefits to the people who live there and to their environment. RMI advocate a seven-step approach as indicated below:

1 recruit people representing a cross-section of community interests;
2 develop a vision of the community's future;
3 identify all kinds of resources available locally;
4 discover the opportunities;
5 use the mutual learning to generate project ideas;
6 evaluate and select specific projects and actions; and
7 implement projects and monitor outcomes against desired goals.

The first aim of economic renewal is to stop the flow of money outside the local community. For example buying from a local retailer is preferable to buying from a large chain because some of the money leaves the community as profit and possibly as wages of some staff. Secondly, buy from local shops so that they thrive and profits are recirculated within the community. Thirdly, new, compatible businesses can be identified to increase the diversity of local shops and facilities, and avoid duplicating those that are already adequately provided. Fourthly, shops and services that are needed can be encouraged to open up in the neighbourhood.

Table 7.3 Four Aims of Sustainable Economic Renewal

- Stop money flowing out of the community.
- Support local businesses.
- Identify compatible businesses needed in the community.
- Encourage new, appropriate businesses to open.

Experience in the USA has shown how to recognise when a community is likely to be ready for economic renewal and when that readiness is absent, summarised in Table 7.4.

Table 7.4 Indications of a community's readiness/lack of readiness for economic renewal

Indications of readiness	Indications of lack of readiness
• A general recognition that something must be done to improve the local economy.	• Overwhelming apathy towards any local concerted action.
• A crisis occurs which brings the community together for example a natural disaster, or closure of major source of employment.	• General faith or expectation that a major change is imminent that will save the town or neighbourhood.
• A few people are already talking seriously about organising a local effort.	• Some event divides the community so that cooperation is not possible.

In order to ensure that any new venture is consistent with the principles of sustainable development it is important to recognise some of the key considerations which help to provide a relevant focus. The first important distinction is that prosperity and improved quality of life can be achieved without growth. Growth just means getting bigger. Focus on size often leads to construction of new buildings and roads which lead in due course to more cars, increased congestion and more pollution. Appropriate development focused on what exists already can lead to re-use of existing buildings, provision of public transport and improving facilities which already exist, thereby avoiding the adverse effects of traditional growth. It may be helpful to use the

Figure X

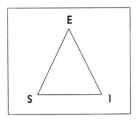

ESI model to focus on what is ethically appropriate for the community, what would be consistent with the community's enlightened self-interest and which innovations from this foundation are relevant and feasible in the neighbourhood?

Creative ideas from other projects include keeping farmers on the land, establishing cooperative businesses, setting up a team to answer queries about opportunities, providing low-interest loans, arranging a recycling scheme, providing outlets for arts and crafts, helping with a literacy programme, tours of local tourist attractions and improving the facilities for recreation.

At some stage it could be helpful to organise members of the community into different groups such as those providing leadership, those offering advice, participants and non-participants, those undertaking particular projects and those who are facilitating others to overcome obstacles. Experience shows that the best results come from maximum involvement and this means recognising how different individuals can contribute willingly from their knowledge, skills and abilities. It is important to have fun, to appreciate achievements and to celebrate success.

Peace Child International

Sometimes it is salutary to look at what children are doing to help the cause of peace and repair the natural environment. One of the most heartening initiatives is Rescue Mission: Planet Earth by the Peace Child International Centre. The way in which the project has involved children from all over the world in writing their excellent publications is enlightening for everyone. A brief description of Agenda 21, the Rescue Mission and Peace Child International is provided in Table 7.5.

Several Chief Executives have admitted that they have got involved with environmental issues because of questions asked by their children or grandchildren. There is no doubt that children can be a potent influence on their parents and that this can lead to interesting initiatives being taken within the organisations that employ them.

Table 7.5: Agenda 21 and Rescue Mission Planet Earth[10]

Agenda 21 was developed over two years by people from 179 countries, leading up to the United Nations Conference on Environment and Development (UNCED) which took place at Rio, Brazil in June 1992. At the Conference, Agenda 21 was agreed and published as a 500 page book with 40 chapters setting out the agenda to achieve Sustainable Development worldwide in the twenty-first century.

Rescue Mission – Planet Earth is Peace Child's flagship project for the 1990s. Peace Child is a non-profit educational charity registered in England, the USA, The Netherlands, Russia and Israel with 30 affiliated organisations. Rescue Mission began by publishing a version of Agenda 21 'that ordinary people can understand', written by 28 children from 21 countries. As they say 'we're tired of seeing our beautiful planet polluted, tired of senseless wars, of the poor getting poorer day by day, of waiting for politicians to make decisions they should have made long, long ago.' The Rescue Mission is a wake up call from children who are equipping themselves to monitor how Agenda 21 is being implemented.

Figure X

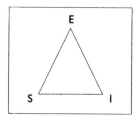

Peace Child and the Rescue Mission is a good example of children recognising the significance of enlightened self-interest, adopting an ethical approach and using innovations to embark on a partnership with adults to inspire people all over the world to take action. They make it look easy! The confidence and creativity of youth is a vital asset which should be encouraged.

LOCAL AGENDA 21

Surveys have been conducted by the Local Government Management Board (LGMB) in 1995 and 1996 to find out the progress being made with Local Agenda 21 in the UK. In the UK the government have placed considerable emphasis on local authorities implementing Local Agenda 21 and most of them have responded. Some of the key points from the 1996 survey are summarised in Table 7.6.

Table 7.6 Local Agenda 21: Progress in the UK[11]

- Local authorities in many parts of the UK were undergoing reorganisation when the 1996 survey was conducted, there are now 542, and 275 responded.
- Of those who responded 90 per cent are committed to Local Agenda 21(LA 21).
- 40 per cent are committed to producing a strategy document by the end of 1996, 14 per cent by some later date and 23 per cent intend to do so but details are not yet decided.
- 75 per cent of local authorities have added responsibilities to the work of existing officers, 34 per cent have appointed new staff and 10 per cent have seconded staff to LA 21.
- 88 per cent state that LA 21 has been discussed at council committees, the most typical committees are 'environment' and policy/resources. Some have involved two or more committees.
- Some form of environmental management system is being adopted by 42 per cent of responding authorities and a further 31 per cent are considering which system to use.
- The integration of sustainability principles in local authority activities is being achieved most readily with environmental services, waste management, land use planning, energy management, transport strategies and green housekeeping, but least well with investment strategies.
- Public awareness is being raised in several ways, especially through environmental education, special awareness raising events, visits, talks, support for voluntary groups and publications.
- Local authorities are forming partnerships with businesses (22 per cent), academia (19 per cent) and non-governmental organisations (27 per cent).
- 37 per cent of local authorities are working on indicators for sustainable development and the most popular theme areas are: resource use, limiting pollution, beauty/distinctiveness, biodiversity, empowerment and protecting health.
- 29 different kinds of new training scheme were requested by 72 authorities with training for policy makers/members/senior officers (16 authorities), training in consensus building/mediation (12 authorities) and training for Environment/LA 21 coordinators (9 authorities) being most popular. Case studies as learning aids and to provide guidance were requested by 11 authorities.

Some of the work being done by local authorities in the UK is high quality, methodical and innovative. Some local authorities have made considerable progress towards meeting stated goals in areas such as reduction in waste sent to landfill and incineration by setting up recovery and recycling schemes. A few examples will give a flavour of what is being done.

Croydon

By May 1996 Croydon Local Agenda 21 had been published as a Consultation Draft.[12] Some 5000 residents of the borough, including business people, councillors, council officers and volunteers had been involved in creating this first document. The project starts by concentrating on the environmental aspects but there is recognition that social and economic aspects of sustainable development need to be included.

The vision of the future describes a strong community with people responsible to each other; access to others electronically as well as through travel which includes bicycles and public transport; lifestyles that are sustainable; an economy providing a greater variety of working patterns which include more people caring for others as well as the natural environment; homes equipped to make better use of resources and congenial places to live with more working from home; more food will be produced within this country and include a higher proportion of protein from vegetable sources; open spaces will provide habitat for wildlife and facilities for a vibrant local community; and waste will be greatly reduced from current levels through re-use and recycling. Juxtaposed with the positive vision is the nightmare scenario which vividly describes all the expected horrors.

The aims are clearly set out in respect of topics such as consumption levels, pollution, waste, diversity of species, and individual awareness of a healthy environment. The report then describes the broad parameters of the natural environment such as air, water, land, public health, the natural environment, waste, energy, transport and the economy. In each case objectives are declared, indicators are identified and targets set.

This is followed by the contribution that can be made by different sectors of the community. A series of pledges are stated with reasons. The whole report is designed to test the degree to which local residents are prepared to agree the aims and contribute to their achievement. The processes and time frames for moving the plan forward are clearly set out and the people involved are named. The document concludes with a comprehensive action plan.

Devon County Council

Devon County Council sponsored a two-day workshop for 30 people, including a county councillor, local people, council office staff, business people and interested parties, in April 1995. They prepared a manual to help run that event which has subsequently been edited and is now available as *A Field Workers' Manual.*[13] This is being used to guide other events in the county and elsewhere to enable people to activate sustainable development strategies for themselves.

The two central themes are Community Development and Designing for Sustainability both of which are key components of Local Agenda 21. A summary of the guidelines given below:

- providing effective learning for people – engaging, reflecting, new thinking and designing ways of implementing ideas;
- achieving transitions – overcoming fear, establishing support networks, following the path of least resistance, and using permaculture ideas (Earthcare, Peoplecare and Fairshares) as the basis for sustainability;[14]
- tackling culture change – recognising the prevailing culture, appreciating the community, encouraging self-reliance, fostering participative leadership, listening to honest feedback, involving people in appropriate ways to design future visions and overcoming difficulties;
- introducing systems thinking in simple but effective ways – recognising the difference between unsustainable high technology and the sustainability of natural systems and permaculture;
- helping people put their own house in order on three key fronts – using fewer non-renewable resources, establishing an edible landscape and purchasing commodities from ethical traders;
- working with nature not against it – every type of landscape is productive in its own terms, biological strategies are the first priority and capable of the highest sustainable yields;
- developing the required skills – building readiness, working with support, observing and listening, thinking and communicating, using inclusive methods, visioning and supporting others.

Other Local Agenda 21 Initiatives

Several other examples of Local Agenda 21 in action are known and a brief mention of some that are particularly interesting is appropriate. Canterbury is a good example of creative partnership developments with industry. Gloucestershire has developed Vision 21 using imaginative methods. Greenwich has established four pilot schemes involving local businesses with help from WWF and NatWest Bank. Sutton Borough Council has earned an enviable reputation and used future search conferences to develop community understanding and commitment. Wiltshire County Council has undertaken a large-scale training intiative using their own adapted version of *The Environment at Work*.[15]

A useful booklet has been produced jointly by Countryside Commission, English Heritage and English Nature to help people implement Local Agenda 21. It includes an explanation of Local Agenda 21, practical tips and case studies and is a well produced guide with good illustrations.[16]

Thousands of communities around the world are now involved in assessing their lifestyles, developing ideas about the future and devising ways of creating sustainable societies in the twenty-first century. Local Agenda 21 has become a global initiative.

CONCLUSION

Many community initiatives are being taken in different parts of the world and only a few have been described. The examples of industrial ecology in Denmark, Austria and the USA provide encouragement that much can be done to improve the natural environment, be compatible with a healthy economy and contribute to social development.

The remarkable achievements in Curitiba, Brazil have been described. An indication has been given of how economic renewal is being tackled in the USA. The Peace Child initiative and their Rescue Mission is an impressive way in which children from many countries are involved, with indicators devised by themselves, in assessing their own neighbourhoods. They have written their own material and devised their own indicators to assess the environment with minimal help from adults.

Local Agenda 21 schemes are spreading around the world as more and more local communities get involved with their own visioning about the future. The creative ways in which these plans are being developed provides hope for the future. Most of these initiatives are involving people in describing their visions, appreciating how social, economic and ecological development need to go together and appreciating what it takes to establish sustainable communities. The next few years will be crucially important as the planning is implemented.

PART 3 MANAGING,
MEASURING AND INITIATING

Part 1 examined 'Our Changing World'. Part 2 looked at 'Three Approaches' that organisations are taking to respond to the ecological, social and economic predicament that faces the world community. Some of the important changes that are taking place have been considered. The awareness of these trends is growing and the response is now more consistent and more widespread. It is coming from businesses, from professional bodies, from local communities as well as national governments and international bodies. The learning curve is still steep and will remain this way for some time as more is learned about appropriate responses to the challenge. There are still many organisations in the business and public sectors that are not involved. Everyone and every organisation can make a contribution to solutions.

Part 3 considers 'Managing, Measuring and Initiating'. These three topics turn the spotlight on the processes for implementation, monitoring progress and taking appropriate, effective leadership initiatives.

Chapter 8 considers the management systems that have evolved to help manage the environmental challenge. These methods, not surprisingly, focus on reducing environmental damage. Environmental management systems are not designed to cover the social and economic aspects of sustainable development. Furthermore a management system that aims to reduce environmental impact may not contribute much to business strategy. Organisations are coping with this in different ways. Some of these approaches are described in this chapter. The importance of a good management system is emphasised by all who are involved. At this stage no single best method can be recommended, but some of the options are considered. The scope for creative developments which integrate the economic, social, and ecological aspects of management in pursuit of the goal of sustainable development is immense.

Chapter 9 looks at the ways in which new measures of performance are being developed to enable organisations to assess how they are doing. Some important examples are outlined and their value discussed. The chapter concludes with some recommended principles and a framework for Corporate Environmental Reporting which has general application.

Chapter 10 considers initiatives that are being or could be taken to help accelerate the advent of sustainable development. Rather than make recommendations or offer prescriptions the intention is to put forward suggestions and options from which a selection can be made to suit particular situations, types of organisation and individual preferences. The concluding section considers the type of leadership that is required when simplistic technical solutions are not appropriate. In these circumstances the effective leader involves, encourages and enthuses people to find relevant solutions. These leaders recognise the importance of shared visions to which people are committed and will implement. When applied within the enterprise it becomes possible to build an organisation that will last; when applied within a community it creates the potential for sustainable societies.

8 MANAGING ENVIRONMENT
AND SUSTAINABILITY

When an environmental issue first emerges a typical response is to deal with it in an ad hoc way. Often this means finding a technical solution, which may require a project if the problem is complex or persistent. In due course as other problems arise the realisation dawns that environmental issues penetrate most aspects of a business. At this stage a systematic approach to environmental management is required and typically an environmental manager is appointed.

Environmental managers can work for a long time, sometimes for years, on a series of projects without moving beyond an ad hoc approach. However, many have found that they earn credibility for themselves and greater benefits for their organisation when some form of systematic environmental management is introduced. As this becomes established it plays an increasingly large part in the affairs of the business and the environmental manager gains more credibility. With this increased credibility environmental managers are more likely to be included in decisions about how to integrate environmental policy into business strategy. Without this ionvolvement any review of business strategy is omitting a major factor in future business performance. During this review it is important to make direct links between the management of environmental issues and the financial performance of the business. As this becomes routine procedure the organisation's approach to sustainability becomes feasible. This creates the conditions for building to last, which also requires an investment in education and training in environmental matters for people at all levels.

This chapter looks at how companies manage their environmental responsibilities and the way in which some have been inspired to go beyond a simplistic approach. It also looks at environmental management education and training – an area which receives less attention than it deserves.

SOURCES OF INSPIRATION

Some organisations have found it convenient to add 'environment' to the safety and health function. This applies particularly to businesses where health and safety are vital. The petrochemical industries are a good example but it is often important for the manufacturing function in other industries. In these situations companies have renamed the department Safety, Health and Environment (SHE) – not always in that order. The thinking is that the procedures adopted for health and safety can be applied to environmental management. In many companies the SHE department, however important its function, is only indirectly involved in business strategy. Environment, health and safety issues tend to influence the way in which business strategy is implemented rather than what that strategy should be. The SHE department is very seldom the source of inspiration about strategic direction.

Environmental issues are now becoming so important in many businesses that they need to be part of the agenda when business strategy is discussed. This requires those involved in the debate to appreciate how to strike a balance between environmental issues and business strategy. A vivid example of getting this wrong occurred when Shell tried to dispose of the Brent Spar, a disused oil platform, by sinking it in the Atlantic. Scientists working for Shell and the UK government had agreed that this was the best course of action but the public disagreed. A boycott of Shell's products occured in Europe, especially in Germany. The threat of the loss of business led Shell to abandon this method of disposal. Chris Fay, Chairman of Shell UK said:[1]

The case for offshore disposal of Brent Spar was prepared with professional skill and judgement, and secured all necessary regulatory approvals. But we were clearly insufficiently sensitive to public concerns and the need for wider discussions. We have learned much from that experience.

Shell were clearly inspired by the Brent Spar episode.[2] They have now initiated a wide-scale debate of the options with independent facilitation.[3] They have also reviewed their environmental approach and issued a number of public statements.[4]

For organisations the source of inspiration for rethinking business strategy varies considerably. Electrolux identify three important sources of help.[5] These are ideas from The Natural Step Foundation in Sweden (see Chapter 4 pages 56–58), the clauses of the Business Charter for Sustainable Development published by the International Chamber of Commerce (see Chapter 1, note 30) and the Rio Declaration from the UNCED Conference in Brazil in 1992 (see Chapter1, note 25).

The specific trigger for different organisations varies depending on their unique circumstances and whether the organisation perceives an opportunity for change. Several examples from different companies illustrate how diverse the circumstances and response can be. These are given in Appendix 1.

POSITIONING

The starting point for environmental management varies considerably. A few are inspired, for others it is a gradual process of evolution. This creates a very mixed situation. Organisations are at different stages of environmental awareness (see Chapter 3, Table 3.6, page 47). For a particular company the nature of the industry and the stage of environmental awareness often leads to contrasting perspectives. Even within the same organisation there will be people who view the subject differently. Some people deny the importance of environmental issues or dismiss them as the responsibility of others, although, increasingly companies are accepting the need to have an environment policy. However, this involvement can vary from a search for ad hoc solutions to those who wish to create sustainable enterprise. Four perspectives are shown in Figure 8.1.

The two segments on the left in Figure 8.1 are self-explanatory. The bottom right box describes practical action which has been taken by many organisations. The top right segment describes strategic thinking which still requires most of the action described in the segment below it to make it happen. However, without strategic thinking much of the action falls short of the far reaching changes which are necessary in some industries.

The process does not stop there because it can influence the assessment and training of suppliers, distributors and retailers. There is no point in a company improving a product only to find that its suppliers, distributors or retailers do not understand that the product is designed to perform its primary function *and* safeguard people *and* the environment as well.

Figure 8.1 Four Perspectives[6]

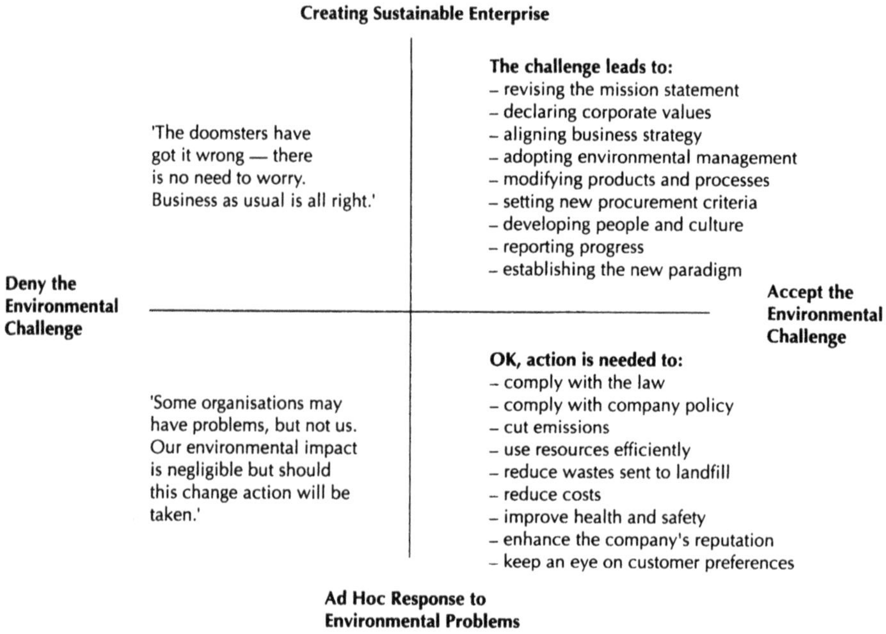

Creating Sustainable Enterprise

'The doomsters have got it wrong — there is no need to worry. Business as usual is all right.'	**The challenge leads to:** – revising the mission statement – declaring corporate values – aligning business strategy – adopting environmental management – modifying products and processes – setting new procurement criteria – developing people and culture – reporting progress – establishing the new paradigm

Deny the Environmental Challenge ——————————————————————— **Accept the Environmental Challenge**

'Some organisations may have problems, but not us. Our environmental impact is negligible but should this change action will be taken.'	**OK, action is needed to:** – comply with the law – comply with company policy – cut emissions – use resources efficiently – reduce wastes sent to landfill – reduce costs – improve health and safety – enhance the company's reputation – keep an eye on customer preferences

Ad Hoc Response to
Environmental Problems

THE SCOPE OF ENVIRONMENTAL MANAGEMENT

'Environment' has many meanings. A few years ago it was confined to the socio-economic market in which business is conducted. Now the emphasis has shifted to the natural environment, while retaining much of its previous meaning. The meaning includes the rich diversity of the natural world, the places where people live and work, where customers use the products or services they buy, and the organisation's impact on the neighbourhood. This means that any environmental management system needs to take account of a wide range of issues.

In manufacturing organisations environmental management requires involvement with the whole process from procurement of raw materials to disposal after use. This includes product specification, purchasing, manufacturing processes, distribution, the product in use, and recovery after use. Design affects every stage of this process and good design can improve energy efficiency, materials selection, use of resources and waste reduction.[7] Organisations which are deeply concerned about environmental matters realise that their entire workforce must be involved in some way. Everyone needs to understand that their contribution can make a difference. To gain the commitment to new ideas from the whole workforce usually requires some form of participative process.

VALUES

An approach favoured by some organisations is to develop a set of organisational values. For example, Kent County Council's (KCC) stated values are:

- targeting those in need;
- 'care in the community';
- caring for the environment;
- strengthening Kent's economy;
- improving quality of life;
- improving local democracy;
- improving the quality of public services;
- providing equal opportunities; and
- listening to and valuing staff.

KCC is still working on implementing these values and seeking ways in which to measure progress.

Rover Cars' declared values are summarised as:[8]

- 'Whole-life' strategy – covering the product from 'cradle-to-grave';
- legislation is the minimum – highest standards are sought;
- best practice – consider environmental aspects even if not covered by law;
- decisions – give priority to environmental factors when all else is equal;
- everyone's responsibility – at all levels of the organisation;
- expertise – commitment to develop environmental capability;
- shared experience – commitment to cross-company sharing of learning;
- openness – keeping the community and authorities informed;
- priority at a breakdown – given to health and the environment; and
- regular monitoring – impact of all processes and their performance.

Many commercial and industrial organisations, as well as local and municipal authorities and professional organisations, are developing their own statement of values and openly declaring them. These statements help focus on both the need for change and the direction it should take. The examination of values is an important first step to be taken before restating corporate purpose. If this is not done the restatement of purpose will remain embedded in the old fashioned, out of date values. A process for reviewing corporate values is described in Chapter 10, Figure 10.1.

An important consideration is the way in which a statement of values is developed. Those involved in its creation understand its meaning and are committed to its intentions. Those not involved often give it scant attention and often discredit it. For large organisations the creation of a values statement to which people throughout the organisation are committed requires a systematic process. Organisations which have done this successfully usually bring people from different departments together and work steadily towards a synthesis of mutually agreed ideas.

Current practices may not conform to the statement of values and this emphasises the need for clear priorities to bring actions into line with intentions. Environmental management emphasises the importance of preventing problems rather than trying to deal with the repercussions. Dealing with pollution is a good example.

CLOSING THE LOOP

If an organisation is to achieve sustainability then one significant goal is to 'close the loop'. Electrolux and Rank Xerox have stated that their goal is to achieve zero, non-biodegradable waste and other organisations have set the same target. They mean avoiding waste, or if that is impossible, recovering all emissions and wastes and re-using components or recycling materials back into the manufacturing process. Trapping these wastes in a filter which is then sent to landfill does not achieve the same result.

In manufacturing industry 'zero waste' includes eliminating the carbon dioxide produced when fossil fuel energy is used for power in the manufacturing process or when equipment is used by customers. This implies an ultimate shift from fossil fuel to renewable energy, but neither manufacturers nor electricity producers have, as yet, made that commitment.

Electrolux are among those organisations which are making real progress, towards sustainability. Their approach is illustrated in Figures 8.2.

Figure 8.2 Vision for a Closed Loop Material Cycle

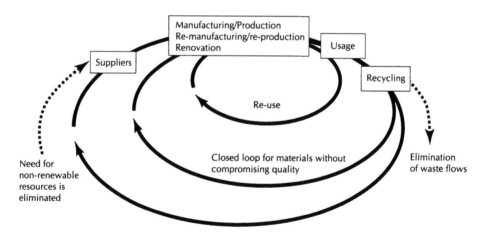

Electrolux have assessed their products in terms of the environmental impact of each product. New product development and the refinement of existing products is thus concentrated on how to integrate new technology to meet customer requirements and environmental considerations. Two examples are given below:

Commercial Refrigerators	Industrial Laundry Equipment
Cooling agents and insulation gases	Energy
Energy	Water
Recycling	Laundry detergent
Noise	Recycling considerations
Working environment	Working environment

Rank Xerox RX show how they dismantle their machines and after inspection re-use certified parts or recycle the materials. This sometimes involves third parties. The disposal goal is to eliminate all wastes sent to landfill. The RX diagram is shown in Figure 8.3.

Figure 8.3 Closing the Loop: Rank Xerox

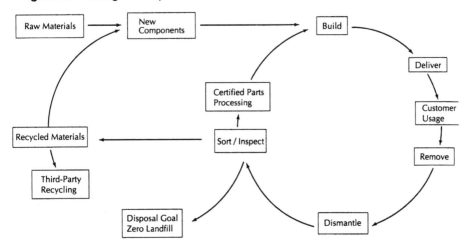

A factor of crucial significance to any business is the scale of environmental impact attributed to each stage of a product's life cycle. This is often summarised in an environmental effects register. In the case of Electrolux the greatest impact comes from the usage of their appliances like washing machines, chain saws, cookers and refrigerators. Some of these are electrical, others, like chain saws, have two-stroke engines and some cookers use gas. To reduce the impact on the environment the efficiency of the product in use is essential. That includes energy use, water usage (where applicable), waste produced during use and wear and tear on the materials.

Procter & Gamble have greatly improved the efficiency of washing powders and liquids. This has reduced the quantities needed to achieve the same washing capability. They have designed and marketed refillable containers thereby reducing the quantity of packaging material and saved money. With these achievements behind them P&G turned their attention to disposal of used packaging materials. They identified other companies whose products produce considerable quantities of packaging which required disposal. This led to the project in Adur District in Sussex, England, supported by ERRA, described in Chapter 7.

P&G published a diagram showing their manufacturing inputs and outputs as shown in Figure 8.4. This illustrates that most of their output is packaged products going for sale and wastes represent only 5 per cent, with 3 per cent going for recycling and 2 per cent for disposal. Of the 2 per cent waste for disposal only 0.1 cent per is solid, hazardous waste.

Figure 8.4 Manufacturing Inputs and Outputs – Procter & Gamble

1995 PROCTER & GAMBLE ENVIRONMENTAL PROGRESS REPORT

Data Section
Introduction

As the diagram below illustrates, P&G's use of material resources is highly efficient. The vast majority of materials are converted to product, and output of manufacturing wastes (which are largely non-hazardous) is low. Data is provided in tonnage for packaging and manufacturing operations. Because air emissions are less than 0.2 per cent these are not broken down into emissions of carbon dioxide, oxides of nitrogen and sulphur dioxide.

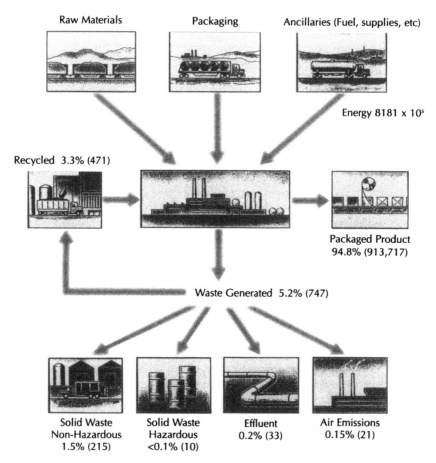

Raw Materials   Packaging   Ancillaries (Fuel, supplies, etc)

Energy 8181 x 10⁵

Recycled 3.3% (471)

Packaged Product
94.8% (913,717)

Waste Generated 5.2% (747)

Solid Waste
Non-Hazardous
1.5% (215)

Solid Waste
Hazardous
<0.1% (10)

Effluent
0.2% (33)

Air Emissions
0.15% (21)

126

P&G have produced a guide to integrated solid waste management and a spreadsheet on packaging life cycles both of which are widely acclaimed.[9]

Other companies have similar goals and achievements but the examples given illustrate what is being done by some of the more progressive organisations. Rank Xerox, Electrolux and P&G have all been involved with environmental matters for some time and they have their own approach to managing quality. A brief look at Total Quality Management (TQM) is helpful because several companies are linking quality and environmental management together.

TOTAL QUALITY MANAGEMENT (TQM)[10]

TQM originally concentrated on engineering quality with the intention of improving customer satisfaction and thereby organisational performance. More recently it has become a philosophy for management whereby attention focuses on customers and enables staff at all levels to contribute ideas towards making improvements. The essence of a good TQM scheme is that employees are motivated and have the authority to introduce changes to improve quality. Where issues cut across departmental responsibilities cross-functional teams are set up to remove the barriers to quality improvement.

The success of TQM has been varied. Many organisations adopt a very procedural approach which can become bureaucratic and leads to much paperwork with the end goal becoming accreditation rather than improved customer satisfaction. A major reason for this is that the implementation of a customer-focused management philosophy is not compatible with the prevailing management style and organisation culture. Organisations seeking accreditation often do not realise that a culture change is required, or believe that they can somehow reap the benefits of TQM without changing their style of management. Culture change can be a daunting prospect and often involves a shift in values. An organisation wishing to improve its approach to quality also needs to gain widespread commitment to a set of values which are clearly stated. Developing a values statement which is widely understood and accepted takes time and effort.[11]

Where TQM has been successful the organisation's culture and values have been developed as part of the programme. The characteristics of TQM include:

- focus on customer satisfaction;
- continuous improvement;
- eliminating the underlying reasons for poor quality;
- recognising that everyone can contribute to achieving quality; and
- calculating the realistic costs and benefits of quality.

ORGANISING ENVIRONMENTAL MANAGEMENT

Organisations that are taking environmental matters seriously have given much thought to the way in which they want to organise their environmental management systems and procedures. The needs of organisations differ depending on the nature of their business, their values and purpose, their size and the stage they have reached in progressing towards sustainability.

Each organisation's environmental impact will influence the way in which environmental management is perceived, positioned and implemented. For example the priorities for a manufacturing company, a chemicals business and a local authority will raise different environmental management issues. For large organisations one of the greatest challenges is to achieve effective integration across departments.

An environmental review carried out with the involvement of employees will help to identify the main issues. Such a review can also be used to prepare management, and define the goals of an environmental audit carried out by external specialists and the preparation of an environmental effects register.

The size of organisation will influence the resources required for environmental management. A large organisation will appoint a manager and probably some additional staff, whereas a small organisation cannot afford this and would need to add environmental responsibilities to someone's job. This inevitably creates the dilemma of how to balance priorities. However, every organisation will need to decide the sort of person who can carry out the role effectively. An environmental specialist can deal with technical issues but may have difficulty in making the links with business strategy. A senior line manager will know the business and have credibility with colleagues but may not have sufficient awareness of the way in which the environmental impact occurs. A team headed by a respected senior line manager working with a small team of specialists may achieve the appropriate balance.

An organisation at the early stages of establishing its environmental management function will need to place emphasis on getting the message across and establishing its role. When environmental management is being formalised there will be a lot of work building commitment and explaining what needs to be done. When this is in place some large organisations find that the central unit can be slimmed down to one or two people with responsibilities delegated to line managers and departmental environmental representatives with cross-functional meetings taking place occasionally.

The ways in which Procter & Gamble, Rank Xerox and Electrolux organise their environmental management functions have been explained in Chapter 5. The Co-operative Bank is in a different situation because its entire business is guided by ethical and environmental policies so the whole organisation is automatically involved in implementing these policies in its day-to-day work. It has decided to encourage this style of banking but it still has an internal ecology unit to keep the bank staff up to date. The Co-operative Bank National Centre for Sustainable Development, established with four Manchester universities, represents one of the largest environmental educational joint ventures in the UK. It is taking the message to a wide range of organisations.

Monitoring Sysems

A prerequisite for monitoring is clarity about objectives, targets and action plans. With these in place it is possible to review results against stated intentions. This is now being done by several organisations and reported in their Corporate Environmental Reports, which are discussed in Chapter 9.

Some organisations introduce interesting innovations to cross-check their performance. The Body Shop and IBM have consulted their stakeholders and published the results. IBM found the reactions to their presentation of their first stakeholder report in 1995 so helpful that they repeated the process in 1996, while the Body Shop have summarised their results on the Internet. [12]

ENVIRONMENTAL MANAGEMENT SYSTEMS

Several different approaches to environmental management have been developed. These include an extension of quality management, a management system linked to environmental auditing and a global initiative taken by a group of multinational companies. The following options are available:

- Total Quality Management (TQM) – or BS5750, ISO9000;
- Environmental Management System (EMS) – or BS7750 (being phased out and replaced by ISO14000);
- Environmental Management and Auditing System (EMAS) – the EU directive (NB BS7750 + audit + a public statement = EMAS);
- International Standards cover seven topics:[13]
  - ISO14000 The Environmental Management System
  - ISO14010 Environmental Auditing
  - ISO14020 Environmental Labelling (Ecolabelling)
  - ISO14030 Environmental Performance Evaluation
  - ISO14040 Life Cycle Assessment
  - ISO14050 Terms and Definitions
  - ISO14060 Environmental Aspects of Product Standards
  - (NB The International Standards are replacing the British Standards); and
- Global Environmental Management Initiative (GEMI).

The British Standards Institute developed BS5750 for managing quality and have developed BS7750 for managing environment. Organisations which have established an effective approach to TQM have used this as the basis for their environmental management. The Rover Group is an example of a successful integration of the two approaches. Prior to the takeover by BMW, Rover Group was run by a Quality Council, with responsibilities similar to a Board of Directors. This located both quality and environment at the top level of decision making where both became central thrusts of their business strategy. Environmental management has the following objectives:

- to comply with or exceed the legal requirements;
- to reduce waste;
- to cut emissions which cause pollution;
- to safeguard the quality and diversity of the natural environment;
- to design sound and effective products and processes;
- to reduce environmental effects from procurement to disposal;
- to integrate environmental management with business strategy;
- to seek continuous improvement;
- to locate environmental responsibility at Board level; and

- to develop and train all staff.

The integration of quality management and environmental management is called Total Quality Environmental Management (TQEM) and is proving both appropriate and successful in a few companies, notably Rover.[14]

The EU have issued a Directive about the implementation of EMAS, which requires reporting by site rather than by organisation. As indicated above this involves an environmental management system, an environmental audit and a public statement which is independently verified. EMAS is being taken up rapidly and over 200 sites were reporting in this way by autumn 1996. The form of the report, which needs to be produced every three years, is left to the discretion of the company adopting the scheme. Several large companies with global systems have difficulty with the emphasis on separate reports for each site. The broad guidelines for the environment statements for each site, include.[15]

- a description of site activities;
- an assessment of all significant environmental issues;
- a summary of emissions, wastes, raw materials, energy and water used;
- other significant factors such as noise;
- a statement of the company's environmental policy, programme and the system implemented at the site;
- the date for the next statement; and
- the name of the accredited environmental verifier.

The interpretation of EMAS by the EU member states varies and this makes consistent implementation very difficult. When this is added to all the difficulties which apply to the effective implementation of TQM it is not surprising that the quality of implementation is mixed. GEMI is a USA-based corporate initiative providing a networking service which is designed to help companies improve their environmental management.

The next few years is likely to herald the next stage of evolution of environmental management to take in sustainability. Three possibilities are:

- SQM – Sustainable Quality Management;
- SMAS – Sustainable Management and Audit Scheme; and
- ISO14070 – Sustainable Enterprise (not yet considered).

In each case it will be essential to provide guidelines which indicate how a company can integrate its business strategy, environment strategy and people strategy into one coherent approach to sustainable development.

Choosing the Right Scheme

The choice of scheme probably makes less difference in outcomes than the way in which it is implemented. However, the following factors may be important in the selection decision:

- Some effective management scheme is already working well. This may be one of the recognised schemes such as TQM, EMAS or ISO9000, but some companies have their own management approach.

- What accreditation is most appropriate – for example European or International?
- Which scheme suits the business best?
- Whether the value added or accreditation perspective is more important?
- Which will cover the business strategy issue more appropriately?
- Whose commitment to the chosen scheme is critical?

PEOPLE AND LEARNING

All these schemes emphasise the importance of training and development for people at all levels. Some companies have made a significant investment of this kind. For example Dow Chemicals took five years to train their people to work in ways which were compatible with their environmental policy. Rover Group ran a seven day environmental management course for 50 of their senior managers who then developed materials for training the next layer of 300 people before cascading it right through their entire workforce of 30,000.[16] Electrolux trained 600 of their people with the aid of The Natural Step in order to create sufficient common ground for the whole organisation to move forward together in a new approach to creativity which linked environmental responsibility with business strategy.

Organisations which have had environmental policies in place for some time know that full effectiveness in implementation depends on people at all levels. They need to understand the policies, how they impact on their work and what they can do to play their part in implementation. A mistake made by some organisations is to rush into action without thinking through the likely implications and alerting management to the importance of follow-up. This is also true of some environmental audits which have been poorly prepared with insufficient consideration being given to the likely outcomes and the follow-up requirements. After the survey has been presented time passes before anything is done about it. This creates doubts about the reason for audit, can lead to questions about the competence of those who approved it and discredits the results. The difficulty of carrying out a successful audit on subsequent occasions is made more difficult.

Preparing a Learning Initiative

No new subject achieves immediate acceptance. It takes time to move from negligible understanding through general levels of competence to mastery, some companies struggle to reach mastery in any subject. In order to appreciate the learning progression some well established subjects can be compared with new subjects. A brief summary of three stages in the process of moving from basic principles to mastery for four contrasting subjects are summarised in Table 8.1. This emphasises the different levels as they might apply to product knowledge, use of information systems, environmental awareness and sustainable development.

Appreciating that these stages occur with every new subject is important and helps to focus attention on how to move from one stage to the next. A carefully prepared plan, which is reviewed regularly, stands the best chance of success.

Table 8.1 From Basic Principles to Mastery

Topic	Basic principles	General competence	Mastery
Product knowledge	Usually well covered	Where communications are effective the levels of competence are kept up to date	Excellent where morale and commitment is high
Use of information systems	As layers are reduced and authority delegated its importance spreads	Depends on quality and effectiveness of training coupled with a willingness to learn	Keeping pace with technological advance is very difficult
Awareness of the environmental challenge	Good where environmental impact is high. Poor in other areas, especially the service sector	Variable – some sectors better than others, some companies take it more seriously. Chief executive commitment is important	A few individuals, scattered across many organisations are very competent, but generally poor
Sustainable development	Recognition of the subject is emerging in a few leading organisations, professional bodies and in local government where Local Agenda 21 is active	The subject is very new but several organisations are now exploring it. It is too early to define competence in a meaningful way	None – the subject is too new, on one has the answer

Assumptions About How to Get the Best From People

The management philosophy in organisations varies considerably. Sometimes it is explicit but often the management style in use can only be deduced from observed behaviour. Many years ago Douglas McGregor identified two contrasting styles of management based on his observations of people at work and named them Theory X and Theory Y.[17] Theory X managers have a pessimistic view of human nature, Theory Y managers an optimistic view.

More recently Tim Hart has described two contrasting sets of assumptions as regressive and progressive in an article on environmental responsibility in organisations.[18] He maintains that those facilitating environmental learning could benefit from being clear about the set of assumptions they are adopting. The assumptions are equally applicable when managing people in any working situation. The reason for taking an interest in these underlying assumptions is that they have a direct effect on how managers behave. An adaptation of these ideas is shown in Table 8.2. The two contrasting approaches are directives and trust.

Table 8.2 Getting the Best Response from People

Directives – people need to be directed and respond best when:
- told what to do;
- given limited, relevant information;
- encouraged by incentives;
- kept in a dependent relationship;
- trained to do particular tasks;
- organised in simple work relationships;
- their minds are focused on their own work or tasks; and
- they are shielded from wider issues.

Trust – people can be trusted and respond best when:
- treated as responsible adults;
- involved in making decisions that affect them;
- encouraged to solve problems;
- interdependence is accepted as normal;
- given opportunities to express ideas *and feelings*;
- allowed scope for self-motivation;
- provided with learning opportunes; and
- given opportunities to participate creatively in wider issues.

Many organisations will say that their operating philosophy does not conform to either of these contrasting lists of assumption – both are too extreme. That is probably true and it may also be true that different managers, within the same organisation, may work from different assumptions. The value in displaying such contrasting beliefs about people is not to encourage the reader to make a choice. There is far more value in writing a set of assumptions applicable to a particular situation then living up to what is written. Even more value can be gained from doing this with work colleagues and establishing a process for giving and receiving feedback on how it works in practice and whether the stated beliefs are being honoured. Thereafter some form of review process can be established to ensure that continuous learning and improvement takes place.

The point about introducing an idea of this kind when considering environmental management and learning about sustainable development is that it has a direct impact on an organisation's performance. Several organisations have found that they can only get the best from their environmental policy when their people accept personal responsibility for playing their part. If they are not treated as responsible when dealing with environmental issues they are unlikely to behave responsibly with other work-related matters. Confronting this issue will not be easy but could be extremely important in many organisations in all sectors.

The Learning Cycle

Appendix 1 shows how different organisations' unique circumstances triggered the decision to take environmental matters more seriously. In each case a top executive had an *insight*. It would be interesting to find out what stimulated the insight. For example some chief executives are known to have been encouraged (provoked?) by their children's questions about their organisation's environmental policy and practices. Questions from a child can be very penetrating, especially first thing in the morning or last thing at night, when vulnerability is often at its greatest. Those are the times when working parents are most likely to see their children.

An insight might take the form of a mental vision, a sudden understanding or wisdom gained from seeing an issue from a different perspective. New insights often act as motivation for doing things differently, or at least to open up a new inquiry. From this there follows knowledge which can be gleaned in a variety of ways. When the insight is confirmed from follow-up investigation the basis for new initiatives has been formed. For many people who have come to appreciate the importance of the environmental challenge and gone on to explore the meaning of sustainable development a learning process of this kind has taken place.

When implementation of the insight, confirmed by knowledge, and influenced by a change in belief, is translated into action, skills are used. Sometimes the required behaviour or skill needs to be developed, sometimes existing skills can be used. A simplified model of the learning process is shown in Figure 8.5.

Figure 8.5 The Learning Process

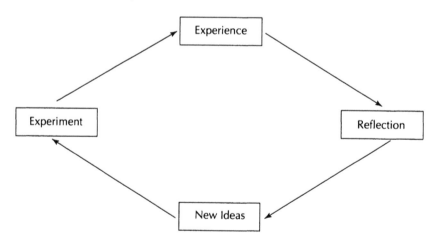

Awareness of a learning process of this kind shown in Figure 8.5 is helpful in order to appreciate the best way to arrange events which allow for people to place different degrees of emphasis on each stage of the learning process.

Environmental learning is a new topic for most organisations and employees. However, it is also important to acknowledge that some organisations have been working in this area for many years. A few, such as Procter & Gamble, have had environmental audits in their plants for 30 years and presenting the findings has provided a rich source of practical information to guide environmental learning.

Learning Methods

The brief explanation about how people glean new insights is important when considering the learning methods to be applied. Anyone who can recall the excitement associated with gaining an insight will realise the difference between that and being taught in a didactic way. There is little doubt which is more powerful and more likely to motivate new behaviour.

The design of learning events enable people to discover new insights for themselves which are usually more powerful and effective. The following learning methods are all helpful in certain circumstances:

- field work – seeing examples of healthy and damaged environments;
- guided research – exploring a topical issue;
- model building – creating a collage or a drawing;
- problem solving – using practical case examples;
- assessing environmental risks – using models for different situations;
- making links with quality – finding out how environment fits into TQM;
- making a video – illustrating an environmental issue;
- role play – experiencing simulated situations in a risk free way; and
- reverse role play – appreciating the perspectives of others.

Whatever methods are chosen to create the right kind of learning opportunity it will always be important to be clear about the objectives of particular events and how the learning can be transferred to the work place. This means:

- articulating specific learning objectives;
- identifying the target population;
- describing the outcomes sought in behavioural terms;
- reviewing results and summarising the learning; and
- developing action plans for use on return to work.

Many managers will be learning new ideas, new relationships and new processes. The following topics are likely to be among those which will increasingly feature for people at all levels:

- understanding the web of life;
- assessing how business activity impacts on the natural environment;
- appreciating The Natural Step's non-negotiable systems conditions;
- applying systemic thinking to solve business problems;
- working with qualitative measures to complement quantitative measures;
- working cooperatively in mixed discipline groups;
- dealing with conflict and using disagreement creatively; and
- recognising the importance of relationships and processes.

The training methods which could prove most effective will go beyond simple instruction. Showing people better ways of working helps to demonstrate that they do work in practice. Involving people in practical work, field work and creative problem solving is even more effective. It also helps to make links with life at home and at work, both of which strengthen understanding. Environmental learning often raises issues which are the source of disagreement.

Dealing with and avoiding disagreement is likely to be a skill that is required when striving for an agreed way to deal with environmental matters. Creating imaginative solutions for the scale and nature of problems that humanity now faces will test the patience and resourcefulness of everyone, but the challenge can also be exciting and fun.

CONCLUSION

The source of inspiration for those who have accepted responsibility for environmental management and the challenge of sustainable development varies considerably. Examples that illustrate how different companies have been stimulated have been described briefly and other examples are given in Appendix 1. Whatever the source of inspiration every organisation is likely to have people with differing views about the extent of the dilemma and its relevance for a particular enterprise. A framework showing four perspectives helps to position particular organisations and the views of people.

Some organisations have reviewed their values and made a public statement about them. This is a powerful and effective approach provided that it is followed up and the values directly influence strategic thinking and implementation. Another effective approach is to close the loops in manufacturing processes, thereby eliminating wastes and emissions. These two very different methods of making real progress are unlikely to be taken until an organisation has developed considerable environmental experience.

Some of the ways in which different organisations are approaching the question of environmental management are described. Some have found that linking environmental management into total quality management is helpful and effective. As the environmental management systems evolve the sophistication will improve. As organisations educate their employees in these matters the role of environmental management units will also change. Some possible future scenarios are described.

The chapter concludes with a discussion about how people learn about the issues, become involved with visions for the future and discover how they can play their part. Some ideas about organising effective learning opportunities are outlined. The final point notes that as more people become involved with the challenge, disagreements are sure to emerge and this will call for creative conflict resolution skills.

9 MEASURING PERFORMANCE
AND REPORTING PROGRESS

There is no doubt that an enormous effort is being made worldwide to investigate how to measure progress towards sustainability. Several organisations have devised their own measures and use these in their Corporate Environmental Reports to describe what they are doing and their achievements.

The UK government's approach, as well as measurements used by some companies for reporting their environmental performance, is used as the basis for this chapter. The simple but powerful framework of non-negotiable systems conditions, devised by The Natural Step, has been used by Electrolux. It can be adapted as a practical basis for reporting performance and progress by drawing on other useful ideas. The result is a framework that can be used by any organisation in their Corporate Environmental Report.

The UK Government's Approach

The UK government has followed up its publication on Britain's Environmental Strategy with annual progress reports.[1] These are complemented by the periodic reports from the Advisory Committee on Business and the Environment (ACBE). ACBE is reconstituted with restated priorities every two or three years and chief executives from different industries are invited to sit on the committee.[2] In addition to these documents the UK government's thinking on Sustainable Development Indicators for the United Kingdom has been published.[3] All these initiatives contribute to the work being done on sustainability. However, many people, especially the environmental groups, argue that the rate of progress leaves a lot to be desired. The absence of coherent sustainable development policies on farming, transport and energy are three areas where criticism is most often directed.

It is now widely recognised that sustainable development requires attention to economic, social and environmental sustainability. The origin of the idea of sustainable development stems from the damage being inflicted on the natural environment and global life support systems. No single company or country can achieve sustainable development in isolation. Nor can environmental sustainability be achieved without linking it to an economic and social framework. Sustainable communities need to be sustainable in terms of economic viability, social justice, and ecological sanity.

The UK government's set of indicators starts with the economy, suggesting that this is still seen by them as the primary issue for a sustainable future. It is also interesting, and disturbing, to note that their definition of sustainable development omits any reference to social justice. Most commentators include social factors within the definition of sustainable development and equity is a crucial consideration, as the widening gap between rich and poor emphasises.[4] In a later section the proposed indicators are compared with other ideas and a proposal for a set of indicators for the future is put forward.

Since the UK government's policies have been determined during a time when one party remained in power for 18 years it is important to consider how another party might deal with this issue. The views of the Labour Party on environmental matters were described by Tony Blair in an address given in February 1996.[5] The following quotations suggest that the Labour Party supports sustainable development, but it is hard to assess how their ideas would be implemented and their draft policy document, now agreed by the Party, pays scant attention to environmental matters:

Labour's vision of a Stakeholder Economy acknowledges a collective obligation to ensure each citizen has a stake in our country's future. . . we all have a stake in the health and integrity of the environment. If there is any case where collective effort is essential, it is in ensuring that our environment is protected so that future generations are not paying the price for our profligacy.

(p3)

The vision set out by Tony Blair includes a three part approach:

1   *To show clear leadership internationally and integrate environmental considerations into decision-making.*
2   *To seek new solutions which promote economic efficiency, link social justice and environmental sustainability together.*
3   *To encourage different forms of community action and local democracy, in order to make progress with concern for the environment.*

(p5)

In this chapter attention is concentrated on the value and use of indicators for public and private sector organisations. However, it is important to be aware of the initiatives being taken by government and whether these might alter should the party in power change. Some ideas which have a more direct bearing on business are the primary focus.

INDICATORS FOR ENTERPRISE

The Centre for Corporate Environmental Management, part of the School of Business, at the University of Huddersfield organised a study of Measures of Sustainability in Business in 1994 and published the outcome.[6] They summarised the results of their study as follows:

A sustainable business would be likely to be operating within an informed, ethical framework which would form part of a wider social ethic. It is likely to have a set of explicit values which represent a bedrock to the way business is done. For many organisations that may represent a paradigm shift towards a more holistic approach. . . A sustainable organisation is also likely to be a learning organisation.

(p3)

The management system they propose for sustainable development is adopted from work done in Canada and is set out in Figure 9.1.

Figure 9.1 A Management System for Sustainable Development

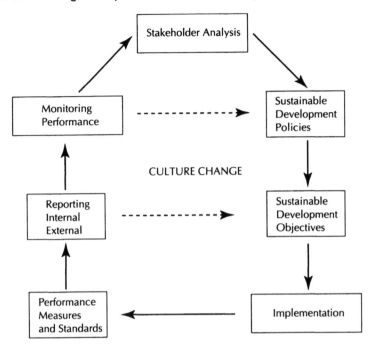

Source International Institute for Sustainable Development

Ashridge Management Research Group and PA Consulting put forward their ideas in 1994.[7] Five drivers of business envirometrics (a term coined by Peter James and Martin Bennett, the authors) are identified. They include financial and non-financial stakeholders, buyers and their own employees, all of whom need a clear sense of strategic direction. The fifth driver is sustainable development and they say:

Sustainable development can only be achieved if business continues to improve its performance radically – a task which requires well-designed measures to identify long-term goals and motivate the continuous improvement which will be necessary to meet them.

(piv)

They go on to identify eight measures:

- impact measures – measures of ultimate environmental impact;
- risk measures – which assess *potential* environmental impact;
- emissions/waste measures – to air, soil and water;
- input measures – to assess the effectiveness of internal business processes;
- resource measures – to record consumption of energy, water, minerals and so on;
- efficiency measures – use of energy and materials;
- customer measures – to register customer satisfaction; and
- financial measures – the costs and benefits of environment-related actions.

Both normalised and aggregate measures are discussed. The former are in two dimensions, for example energy consumption per tonne of output, while the latter convert multi-dimensional measures into a single dimension. They describe the work on environmental performance measures in the oil, petrochemical and paper industries by Deloitte Touche Tohmatsu. Twenty-four different measures have been identified for the paper industry, listed under seven headings, with a footnote that very few, if any, paper companies will be using all these measures. The headings are:

- Materials use
- Energy consumption
- Air emissions
- Soil emissions
- Waste disposal
- Hazard – including fines for non-compliance
- Ecological change – species depletion index

James and Bennett's report is based on actual practices in several pioneering companies in Europe and the USA. Their objectives include categorising various kinds of measures and linking the measures to business strategy. They state that the subject is complex and the brief summary provided above bears this out.

THE WEB OF LIFE

Fritjof Capra, well-known for his books *The Tao of Physics*[8] and *Belonging to the Universe*,[9] has also written *The Web of Life*.[10] *The Web of Life* contains many ideas which are of value for the measurement of performance in the context of sustainable development.

Capra describes the evolution of living systems – organisms, social systems and ecosystems – and the development of systems thinking. During the past few decades the important elements have been identified. He discusses the way in which organisms have been elaborately classified. He explains the importance of cells – the enzymes, proteins, amino acids and so on of molecular biology. He outlines the evolution of systems thinking in biology, gestalt psychology and ecology and the organising relations of living systems. He describes how networks link organisms together in feeding relationships, producing food chains and food cycles which make up the web of life. He explains how systems thinking shifts the perspective from objects to patterns of communication within networks and how positive and negative feedback loops function. The explanations are fascinating but one particular perspective emerges strongly.

The crucial importance of the living cell is given considerable emphasis by Capra when he says: 'Even the simplest living system, a bacterial cell, is a highly complex network involving literally thousands of interdependent chemical reactions'(p26). His theory states that living systems are represented by a pattern of organisation, embodied in a physical structure with an ongoing process. 'These three – pattern, structure and process – are three different but inseparable perspectives on the phenomenon of life' (p27).

Three important issues come out of Capra's analysis:

- the crucial importance of the living cell – this is endorsed by The Natural Step as discussed below;
- the need to recognise the importance of those elements which cannot be quantified because *quality* is the critical issue; and
- patterns and processes need *qualitative* measures.

There is an important link between Capra's work and The Natural Step.

THE NATURAL STEP

The work of The Natural Step has been mentioned several times and a summary is included in Chapter 4. Five statements by the scientists were:

1 Environmental problems need to be tackled at the systemic level.
2 The living cell is concerned only with the conditions necessary for propagating and sustaining life.
3 There is much less difference between the cell of a human and that of a plant than is commonly believed.
4 Humans are not masters of nature but part of it.
5 The living cell, with free energy from the sun, continually produces quality by restructuring waste into valuable resources and propagates and sustains life through its cyclic processes. There is no other comparable production unit on Earth.

The emphasis given to the living cell by both Fritjof Capra and The Natural Step is striking.

It is now time to revisit the non-negotiable systems conditions which the scientists identified, not only as a guiding principle for strategy but also as the basis for sustainable development. The four non-negotiable systems conditions are:

- Substances from the Earth's crust must not systematically increase in nature.
- Substances produced by society must not systematically increase in nature.
- The physical basis for the productivity and diversity of nature must not be systematically diminished.
- There must be fair and efficient use of resources to meet basic human needs everywhere.

The significance of the green cell coupled with the non-negotiable systems conditions lead to an investment strategy for all enterprise, which identifies four goals:

- to reduce our dependence on mining and fossil fuels;
- to reduce our dependence on persistent, unnatural substances;
- to reduce our dependence on depletion of nature; and
- to do more with less.

A Simple Framework

A summary of indicators shows a comparison between what is recommended by The Natural Step, the paper industry (as one industrial example) and the UK government (Table 9.1). The Natural Step does not include the economic dimension. The first three systems conditions emphasise ecology and the fourth moves into the social dimension. The UK government and the paper industry do not include the social dimension. The paper industry includes the economic dimension only in terms of the firm.

Table 9.1: A Comparison of Indicators

The Natural Step	Paper Industry	UK Government
Substances from the Earth's crust must not systematically increase in nature	• Materials use • Energy measures • Mineral extraction	• Transport use • Energy
Substances produced by society must not systematically increase in nature	• Air emissions • Soil emissions • Waste disposal	• Leisure and tourism • Overseas trade • Acid deposition • Air • Waste • Radioactivity
The physical basis of nature must not be systematically depleted	• Hazard measures • Ecological change measures	• Land use, cover and landscapes • Water resources • Forestry • Fish resources • Climate change • Ozone layer depletion • Fresh water quality • Marine • Wildlife and habitats • Soil

We must be fair and efficient so as to meet basic human needs everywhere		• The Economy
		  – Transport use
		  – Leisure and tourism
		  – Overseas trade.

Table 9.1 provides the basis for combining the recommendations and suggesting a simple framework which covers all three dimensions of sustainable development, namely ecological, social and economic. Based on these three dimensions, five guiding principles, based on The Natural Step's non-negotiable systems conditions, can be defined. The consensus of Swedish scientists from different disciplines agreed the first four and one more is added to include the economic dimension.

The result is shown in Table 9.2. The first column restates the three dimensions, the centre column sets out the five guiding principles and the final column suggests what could be monitored – the indicators. The value of this framework is its simplicity. All three dimensions and the guiding principles need to be covered by every organisation. However, the indicators could be selected depending on the nature of the business and local circumstances. Independent verification could then assess three things:

1 adherence to the guiding principles;
2 appropriate selection of indicators; and
3 progress towards stated goals.

This recommendation is proposed to simplify the way in which any organisation can approach the question of monitoring their progress towards sustainability. It also provides a standard framework which is readily understood. There would still be a need to prioritise action and not make sweeping assumptions such as 'all recycling benefits the environment'.

Table 9.2 A Proposed Framework for Indicators

Dimensions	Guiding Principles	What to Monitor
Ecological Sanity	1 Substances from the Earth's crust must not systematically increase in nature	Benign, efficient technology Fossil fuel energy Mineral extraction Renewable energy developed Reuse and recycling Transport use
	2 Substances produced by society must not systematically increase in nature	Acid deposition Air quality Leisure and tourism Overseas trade Waste disposal – zero target
	3 The physical basis of nature must not be systematically depleted	Climate change Fish resources Forestry Fresh water quality Human health Land use, cover and landscapes

		Ocean and coastal waters quality Ozone layer depletion Radioactivity Renewable energy Soil quality Water resources Wildlife and habitats
Social Responsibility	4 There must be fair and efficient use of resources to meet human needs – do more with less	Democracy Equity Peaceful coexistence Security Stable population Sustainable yields
Economic Viability	5 Local, national and global economies must be viable and balanced	Inflation – target: zero Long-term debt – target: zero Increased proportion of local trade

New Economics

New economic thinking is making its mark in several countries. In the UK it is led by The New Economics Foundation (NEF). Their publication *Accounting for Change* provides many insights on the subject of indicators for sustainable development.[11] One of their ideas is to assess indicators in terms of the degree to which they are 'warm' or have resonance. This means that they are widely understood, appeal to people because their meaning is self-evident and they are reasonably accurate. 'Cold' indicators which have little or no resonance may be accurate but they are generally not understood or have an esoteric meaning which is understood only by well-informed people. Their general use is quite limited. The dimensions of accuracy and resonance can generate four different kinds of indicator as shown in Figure 9.2

Accurate indicators stimulate thought, create warm feelings about their significance, have a meaningful context, are generally understood, lead to action and are highly desirable. These characteristics of 'good' indicators can provide a helpful basis for further research. This could be focused on both quantitative measures and qualitative measures, thus picking up a point emphasised by Capra.

The local appropriateness of measures is important. NEF point out that in the UK the destruction of hedgerows gives a good indication of habitat destruction. However, in Chile's capital city, Santiago, this has no meaning, but the number of days when the Andes mountains are visible has a lot of meaning because Santiago people suffer from severe, chronic air pollution. Several people are also experimenting with ways in which environmental quality can be assessed by measuring such things as the mix of species and their numerical strength. Children can play a part in counts of this kind, learn a lot, have fun and provide benchmarks of value to the whole community.

Figure 9.2 Indicators – Resonance and Accuracy

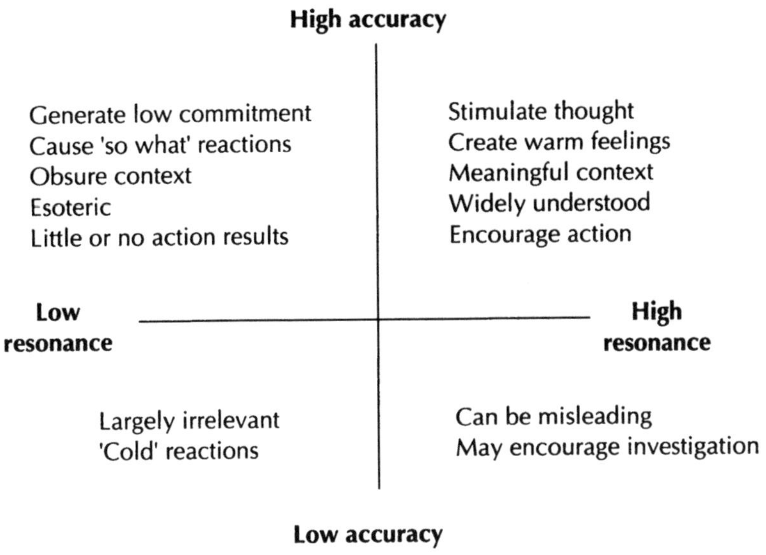

High accuracy

Generate low commitment	Stimulate thought
Cause 'so what' reactions	Create warm feelings
Obsure context	Meaningful context
Esoteric	Widely understood
Little or no action results	Encourage action

Low resonance ———————————— **High resonance**

Largely irrelevant	Can be misleading
'Cold' reactions	May encourage investigation

Low accuracy

Adapted from New Economics Foundation paper 'Accounting for Change'

MEASURING CORPORATE PERFORMANCE

The way in which companies are measuring their corporate environmental performance highlights some good practices which correlate with many of the elements in Table 9.2.

Electrolux

Electrolux have been influenced by the work of The Natural Step and have adopted the four systems conditions (see Chapter 4) as the basis for reporting progress. Electrolux state clearly that they are an industrial company, not a green business but now conduct their business increasingly in accordance with ecological demands. In their 1994 environmental report they use the four headings to indicate how they have made progress during the previous year. Their achievements under these four headings have been listed in Chapter 5, pages 78–81.

Electrolux have worked out their own methods for assessing their production units and facilities. This includes total use of energy and water resources and emissions of carbon dioxide. They provide figures for units of energy used and in relation to added value. These figures are shown in Table 9.3.

Table 9.3 Electrolux Energy Consumption

Year	Number of Units	Per added value KWh/kSEK	Per heated surface kWh/m²	Energy costs % of added value	CO₂/added value kg/kSEK	Water/added value m³/kSEK
1987	91	184	637	3.45	47	1,8
1988	100	168	630	3.15	44	1,5
1989	137	166	634	3.42	45	1,9
1990	150	160	615	3.49	44	1,6
1991	156	156	609	3.54	45	1,5
1992	156	149	609	3.33	43	1,2
1993	165	128	608	3.24	37	1,0
1994	181	112	585	3.08	33	0,7
1995	173	117	587	3.05	35	0,7

* kSEK refers to '000 Swedish Kroner, approximately SEK12 = £1.)

Table 9.3 illustrates the creative way in which one company has devised its own simplified reporting procedure with clear measurements that can be easily understood.

IBM

Brian Whitaker, Environmental Affairs Manager, IBM Northern Europe was very disappointed that the IBM environmental report attracted little attention and comment. A great deal of time and effort had been expended, apparently to no purpose. In order to change this situation he decided to consult stakeholders, employ an independent organisation to carry out the survey and present the results at a public meeting. In September 1995 their report *Consulting the Stakeholder* was presented at a well-attended meeting in London.[12]

The outcome not only provoked much more comment, which was largely favourable, but also established benchmarks on stakeholder views which could be used in future years for comparison. The benchmarks described IBM's environmental performance profile and stakeholders' views on priorities. Some of these priorities had been monitored by IBM, others were new areas. The benefits IBM have derived from this survey are:

1 independent assessment of IBM's environmental performance, judged by the company's stakeholders;

2   reprioritisation of areas for attention; and
3   inclusion of new topics, such as Information Technology (IT) in pursuit of sustainable development.

IBM's list of eleven priorities, show two sets of measures. The first is the score achieved by IBM compared with the best practicable industry performance as defined by ECOTEC, the firm which carried out the survey. The second measure is the IBM performance target. The environmental performance profile is used to identify the targets which have the highest priority, namely:

1   environmental management;
2   IT in pursuit of sustainable development;
3   IBM's product stewardship; and
4   environmental aspects of IBM's customer relations.

The report also indicates the areas in which IBM is deemed by stakeholders to fall short of best practice, which are:

1   transport, where stakeholders want IBM to minimise the environmental impact of transport used by employees and to move products. They also want IBM to develop and supply technology which helps IT users reduce their own transport-based environmental impacts; and
2   commercial activities, where stakeholders want measurement and reporting of the environmental impacts of IBM's non-manufacturing, office based activities.

Social Accounting at Traidcraft

Traidcraft is a Christian trading company bringing goods from farmers and crafts men and women in 26 developing countries to the UK market. Their foundation principles have a strong ethical emphasis which state their intention to establish a trading system which 'expresses the principles of love and justice' in partnership with their suppliers. In order to find out if they were living up to their stated intentions Traidcraft undertook a social audit with the help of the New Economics Foundation (NEF).

Traidcraft have published their *Social Accounts* for 1994–1995 and will repeat the process annually, with refinements based on their learning.[13] Traidcraft's purchases from developing world countries grew by over 30 per cent from 1990 to 1995. Total sales have risen from £1 million in 1983 to nearly £7 million in 1995. They are still trying to reduce their accumulated deficit which has reduced from £180,000 in 1992 to £75,000 in 1995. The report covers five topics:

* key performance indicators;
* producer perspectives;
* review of craft sourcing;
* a new food sourcing policy; and
* what Traidcraft learned from the survey.

The performance indicators used by Traidcraft have been established for some time. They include volume increases (25 per cent over five years), percentage of total sales of goods from the developing world (67 per cent) and continuous growth in revenue to their partners (suppliers) in developing countries.

For the 1994/95 report Traidcraft concentrated on Bangladesh. Their producer perspectives include continuity of relationships, product development, avoiding any producer becoming unduly dependent on Traidcraft, paying fair prices, prompt payment for goods, providing feedback on product quality, taking environmental matters into account and providing social benefits to individuals and the community. The report provides several stories about the actual situation of named suppliers, as a background to the report.

For example the Chandpur Cottage Industries (CCI) provide orders for jute handicrafts for 58 women's groups with 1200 members. Their earnings depend on the hours worked and many of the women have other sources of income as well as household responsibilities. Saida, a group leader, has worked in women's groups linked to CCI for 16 years and now has her own group. Part of Traidcraft's mission is to help their suppliers so it is important to assess how Saida benefits from being a Traidcraft supplier. Her earnings ensure that her four children get proper schooling, and she has learnt to read and write which enables her to understand the legal rights of women. She has reroofed her house from her savings. Saida's group decide together who will join the group but only admit new members when there is sufficient work.

The craft sourcing policy aims to address inconsistencies in applying purchasing criteria and to strengthen the supplier base through a rational programme of sourcing and support. The food sourcing policy aims to apply new principles which include: sustainable agriculture, maximising the value added, avoid compromising local food security and may consider targeting specific groups who have been marginalised.

The *Social Accounts* report covered the views of different stakeholder groups on how Traidcraft is perceived, these included:

- customers perspectives, especially mail order;
- retail and wholesale customers;
- public perspectives;
- employees perspectives; and
- shareholders perspectives.

All these stakeholders have opportunities to comment across a wide range of issues and, in general, Traidcraft are highly regarded. However, the overall outcome of the social audit indicated some interesting areas for Traidcraft to improve, namely to:

- more closely examine customer perspectives on product and service quality;
- pay more attention to the impact of wider development education;
- consider a consultation process with other bodies such as the European Fair Trade Association and UK fair trade organisations;
- clarify the criteria for producer group selection;
- strengthen the indicators for environmental performance; and
- consider integrating social bookkeeping into its information system.

Traidcraft has embarked on an approach to social accounting which is both ambitious and creative. They have helped to form the International Association for Social and Ethical Assessment, with a steering group set up in 1995. They are already attracting enquiries from well-known larger organisations wishing to consider social accounting for themselves.[14] NEF, who worked with Traidcraft, have published their own Social

Statement and publish periodic papers on the subject of social audits, ethical accounting statements and social statements.[15]

Smaller Firms

There is no doubt that improving social and environmental performance is different for large and smaller firms. The large firm can appoint a competent person to examine the issues and make recommendations. The cost in relation to other overheads, service functions, or investments in future strategies is small, even negligible. For the smaller firm this is not the case. If an initiative is to be taken it will usually mean that one of the directors agrees to look at the issues, alongside other responsibilities. The dilemma of balancing urgent and important priorities becomes a real, immediate problem. There is no easy answer to this situation. However, three things can be done which would help the smaller firm:

- having a readily understood framework for assessing sustainability;
- finding two or three low-cost initiatives which pay for themselves quickly; and
- using early successes and/or savings for more ambitious initiatives.

The guiding principles summarised in Table 9.2 may be a good starting point. These provide a clear basis for looking at any business and could, in many cases suggest a few early initiatives which can be taken with relative ease. Alternatively the criteria shown in Figure 9.2 may provide the basis for a smaller firm to generate its own, company-specific sustainability indicators. The first move is often the hardest, thereafter progress can be made by moving forward steadily. At some point there may need to be a more radical shift. By that time several people in the firm should be aware of the issues and accept the need for change. The prospects for getting commitment to strategic renewal will be easier when there are strong external pressures as well as the internal insight. External pressure may come from a larger organisation which is struggling to meet its own agenda and applying pressure on smaller firms which are its suppliers.

Another approach, recommended by Welford and Jones,[16] is for small firms to collaborate, make alliances with local authorities and move forward together with new initiatives. They also point out that in some cases the smaller business, being less cumbersome and more versatile, is well placed to set an example and take the lead. There are advantages in being small, nimble and creative.

CORPORATE ENVIRONMENTAL REPORTS (CERs)

Various guidelines have been proposed for the presentation of environmental reports. One example comes from the Public Environmental Reporting Initiative (PERI).[17] Their guidelines spell out what is recommended under the following headings, but leave the selection to individual firms:

1   Organisational Profile
2   Environmental Policy
3   Environmental Management
4   Environmental Releases

5   Resource Conservation
6   Environmental Risk Management
7   Environmental Compliance
8   Product Stewardship
9   Employee Recognition
10 Stakeholder Involvement

Several assessments of environmental reports have been published.[18] *Tomorrow* magazine pays particular attention to their quality, what they cover, how they are verified and how the reporting process is likely to develop. This will depend on the degree to which companies succeed in integrating their environmental policies with their business strategy and demonstrate that they are moving forward in a coherent way. It will also depend on their ability to show how their combined business strategy and environmental management influence their financial performance. The latter is seldom apparent in current environmental reports with the result that financial institutions pay little attention to Corporate Environmental Reports.

For the future CERs are likely to become more valuable, more open, include the views of all stakeholders and be independently verified. A summary of developments might include the following:[19]

- most companies will publish a CER;
- the format will be simpler and they will be easier to read and understand;
- progress towards sustainability will be reported with quantitative measures, qualitative assessments and stakeholders' views;
- a standard format, possibly based on or including The Natural Step systems conditions, will emerge;
- the interaction of environmental policy, business strategy and financial performance will be more explicitly stated; and
- social and environmental auditing will complement financial auditing with independent verification.

CONCLUSION

The search for indicators of sustainable development is taking place worldwide involving governments, public bodies and academics. However, if mindsets remain largely unchanged the focus will confirm present values and orientation. Where proposed indicators omit key aspects of sustainable development, such as the social dimension, they are unlikely to be adequate. However, by bringing together the ideas about indicators from different sources a more comprehensive approach can be developed.

Organisations like the Centre for Corporate Environmental Management, at the University of Huddersfield and Ashridge Management College have put forward their views, both of which provide helpful ideas. Fritjof Capra has added to his already substantial contribution to environmental management with a fresh examination of the web of life. The Natural Step, a Swedish initiative, has developed a useful set of non-negotiable systems conditions, backed by a consensus of scientists. Capra and The Natural Step emphasise the importance of the living cell – the only truly productive resource on Earth.

These ideas are combined to propose five guiding principles for sustainable development. They are easy to understand and this makes them a valuable basis for a standard approach, which can be used by any organisation. The fact that Electrolux, Ikea and the Co-operative Bank are developing their future business strategy along the lines proposed by The Natural Step demonstrates that this is both feasible and makes business sense.

IBM and Traidcraft have developed their own unique ways of verifying the appropriateness of their policies by doing stakeholder surveys. IBM call this 'consulting the stakeholder', Traidcraft call it 'social accounting', but there is a considerable overlap in the reasons for doing these surveys. In both cases independent organisations are used in addition to internal resources.

The question of Corporate Environmental Reports (CERs) is considered and a framework suggested based on the work of the Public Environmental Reporting Initiative (PERI). How these reports might develop in the future is summarised using guidelines suggested by John Elkington.

The choice of measuring and reporting methods rests with each individual firm, but their reports are scrutinised and compared with others, thus driving for higher standards. As the measures and the reporting processes become more coherent there is likely to be some convergence towards a standard approach. Until that happens the most crucial factor will be the commitment of top executives to the indicators, the reporting style adopted and the use made of the findings by each individual organisation.

10 TAKING
INITIATIVES

The reasons why further significant changes are needed from governments, businesses and individuals have been covered in earlier chapters. This chapter describes some practical ideas, processes and methods which work and need more widespread application from businesses of every kind. The ideas are still being developed so they are proposed as options or a springboard for further creativity and innovation. There are many opportunities for leaders at every level to take initiatives and move forward the search for solutions.

PRACTICAL METHODS

Business success is dependent on several important factors. These include clarity of purpose and values, consistency of business strategy, adaptable organisation culture and strong customer focus, with committed leadership.[1] The methods used by successful companies and how they convert general guidelines into practical action are worth careful examination. However, every company will need its own unique approach rather than trying to replicate what others have done. In a turbulent social, economic and ecological environment it is equally important to look at how successful companies identify the context and timing for major change initiatives. In the coming years many companies will struggle with how to build an enterprise that will last. A few are already involved with the early stages of this process and it is worth looking at what they are doing.

For example Dr George Howarth makes a particular point of emphasising that the environmental initiatives taken by Smith & Nephew are always linked to a topical business context.[2] Without this explicit link it is very hard to gain the attention of line managers who have other priorities.

Dr Allan Kupcis, CEO of Ontario Hydro in Canada, introducing their Sustainable Development Report 1995, sets out five sustainable development indicators:[3]

- environmental integrity – complying with the law and moving beyond this;
- increasing energy and resource use efficiency;
- renewable energy – increasing development and use;
- financial integrity – asset utilisation and operational efficiency; and
- social integrity – safety at work, equity and community interests.

Ontario Hydro set up a Sustainable Energy Development Task Force which reported in 1993 with 98 recommendations. Steady progress has been made since then with clear policy statements, financial and environmental targets, action plans, management strategies and a revised reporting framework. Their report sets out the strategy and achievements against the five sustainable development indicators.

To be effective, initiatives such as those taken by Smith & Nephew and Ontario Hydro, require understanding and commitment to policy statements, strategic goals and management methods at every level in the organisation. This is best developed through debate, exchange of ideas and broad agreement to the way forward.

Successful methods for doing this are being refined through increased demand from organisations of every kind.

TASK FORCES

These are a practical way in which complex problems can be studied in some depth and a report submitted. Ontario Hydro is a good example of using a task force to look at sustainable development. Clear terms of reference are important and the composition of the task force needs to be chosen with care so their report carries credibility. Even quite small organisations can set up a task force of two people to study an issues and either make a recommendation or set out options for a Board decision. When this involves further substantial debate an appropriate meeting, or series of meetings, which can do justice to the report's findings, need to be arranged.

AWAYDAYS

Large companies like Shell, BP and ICI have used the idea of 'awaydays' for many years. These involve taking a group of senior managers away for a day or two to give attention to specific issues which it is very difficult to deal with effectively in any other way. Regular meetings tend to be dominated by short-term operational matters and the important strategic issues frequently get inadequate attention. Awaydays are a good way to have a thorough discussion about important matters, to exchange views and to develop common understanding which leads to agreed policies, strategies and effective action.

Organisations which use awaydays regularly know that they work best when the preparation for them is meticulous and the briefing of those who will take part is thorough. This saves a lot of time at the meeting itself because everyone arrives knowing the context and the purpose and having prepared for it. A disadvantage of this approach is that it can sometimes result in established positions being reinforced. This may make it more difficult to reach mutually agreed outcomes but when they are reached there is much greater confidence that divergent views have been considered and the resulting agreement is more robust as a consequence. If agreement proves illusive a more open exploration of vision and values may be needed.

VISION AND VALUES

Visioning is one of the most powerful methods for understanding how things might be in the future and has the advantage of building commitment to make change happen. The process works best when it is managed by skilled facilitators. Major changes in large organisations involve many people at different levels. To meet the growing demand for larger meetings which are effective, new methods have been developed and are being refined. It is now increasingly feasible to bring together several dozen, even several hundred people together to take part in a large meeting and achieve a successful outcome. In contrast with the awayday approach these meetings often work best when people arrive with an open mind. It is still important to set the context and purpose and create the right expectations about how the meeting will be conducted.

There is a growing body of knowledge and expertise about working creatively with large groups of people. The processes involved are being applied in several areas. These might be to establish a common vision, to review corporate strategy or to explore emerging values. In each case the meeting is designed to build commitment from which effective action can flow.

The methods are appropriate for businesses but are also used in the public sector and by Local Agenda 21 groups. A useful summary of different approaches has been compiled by The Centre for Large Group Interventions (CLGI).[4] A comparison of three large group methods is summarised in Table 10.1.

Table 10.1 Large Group Methods Compared

Future Search Conferences
The primary purpose is strategic planning and the method helps groups, which may be in conflict, to find common ground. It is suitable for 12 to 64 people who are together for two to three days, often in small groups of six to eight people. A skilled facilitator, working with co-facilitators for larger groups is essential. The process was developed by Marvin Weisbord in the USA.

Real Time Strategy Change
The primary purpose is to help achieve organisational change by creating and aligning mission, vision and strategy. It is suitable for 12 to 200 people and takes two to three days. Much of the time is spent in groups of eight in breakout rooms which then come together in one large room. Skilled facilitation is essential with the number of facilitators dependent on the overall size of the group. The method was developed by Kathie Dannemiller and Robert Jacobs in the USA.

Open Space Technology
The purpose is to achieve a high energy, action-orientated meeting and resolve underlying complex differences from which vital issues and strategic ideas emerge. It is suitable for 10 to 1000 people and takes three hours to three + days. A large room, accommodating all participants in a circle or concentric circles is required, with breakout rooms for work in smaller groups on topics decided during the session. Skilled facilitation is essential. The method was developed by Harrison Owen in the USA.

Source Martin Leith, CLGI Guide, 1996

Some form of large group intervention could be very helpful for exploring other topical issues. For example exploring the meaning and implications of the 'inclusive approach', as recommended by the RSA report on Tomorrow's Company. The inclusive approach means recognising accountability to all stakeholders, such as shareholders, customers, suppliers and employees. Some businesses are adopting or exploring these ideas.[5]

If an exploration of corporate values seems appropriate but a self-administered approach is preferred the process set out in Figure 10.1 can be used or adapted.

Figure 10.1 Exploring Company Values

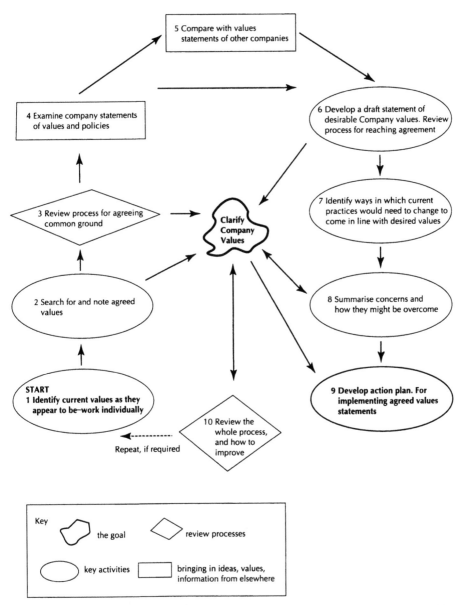

Source: Based on 'An Exploration of Values,' contained in *Vitality and Renewal* by Colin Hutchinson, published by Adamantine, 1995, p203.

CHANGING MINDSETS

One of the most important reasons for considering the large group meeting or a carefully planned awayday is that these approaches are often the best way to test whether well established mindsets are still valid in a world that has changed. It is not always obvious that they need to be changed because established business practices which have worked well for many years may seem satisfactory. To challenge successful methods may seem ridiculous but the real issue is whether these methods will remain appropriate for the future. At times of rapid change it is prudent to examine the basis of today's success quite closely. Failure to do so may mean that a decline in business fortunes becomes unstoppable. Judging the right time for a review is not easy.

The S-shaped curve illustrates the dilemma very well. When results indicate a rising trend and everything seems to be going well there is little justification for change. When a rising trend ceases and a decline sets in it is very hard to recover. The best time to undertake change is often when the growth curve becomes less steep and there are signs of flattening. At this time all the signals will still be saying 'all is well' but that is when a new direction is most likely to be accomplished successfully.[6]

Many senior executives say they have too little time to think and the result is that companies tend to perpetuate current practices, defer change and perpetuate 'business as usual'. It may not seem like this because alongside the feeling of 'not enough time to think' is an equally powerful feeling of constant change. No sooner has one change been introduced than another comes along before the first is properly established. Both feelings of 'no time to think' and 'constant change' can block consideration of major issues such as exploring values, adopting the inclusive approach or applying the principles of sustainability. However, their importance is recognised by companies like Electrolux, Ontario Hydro and Procter & Gamble as the basis for future competitiveness. Companies like these are at the forefront of building to last – a challenge likely to be taken up by more companies in the coming years.

The methods described above may seem time consuming and inappropriate but they are proving successful for many organisations. With careful planning they work well and often surprise those who take part.

THE SMALLER BUSINESS

Smaller businesses face a difficult dilemma.[7] In a large organisation the top executives can take time to consider major strategic issues because that is their prime function. The money earning activity is the responsibility of others and will continue even when the whole board is away for two or three days. In the smaller business this is not the case – the directors are more directly involved in bringing in the earnings. Finding the time to be away is never easy, but it is especially difficult for smaller organisations when it requires the presence of the whole board. Despite this it is often in their interests to get away together for a few hours or a day or two even if it means doing this out of normal working time. In fact many forward thinking smaller business's do just that.

Large organisations find it easier to give someone a special assignment to look into a new topic, such as 'the implications for our business of adopting a built to last approach'. In the small business this probably needs to be done by one or two people taking it on in addition to their normal role or paying for an external study. The internal approach is often preferred because the learning that takes place is kept within the business. The extra demand on time is accepted when those who take it on are part owners of the business with a strong interest in both short-term results and lasting success.

At the implementation stage the smaller business has an advantage over the big organisation because fewer people are involved in understanding and applying new strategies and this process can be undertaken quite fast.

Another approach adopted by some organisations to challenge mindsets is to expose senior executives to radical thinking by involving people from NGOs and campaigning organisations.

WORKING WITH 'CAMPAIGNERS'

In order to explore topics which are not part of the traditional business several companies seek help from those who have campaigned against them. Environmental groups and human rights organisations are among those which have been involved in this way. However, choosing the right source of help is not easy.

Voluntary groups have different strengths. Some are good at campaigning, others at working collaboratively. Some concentrate on revealing problems others are now working on solutions. Selecting the right kind of voluntary organisation, or the right people from that organisation, requires care. Some of the orientations are described in Figure 10.2.

Figure 10.2 NGOs and Voluntary Organisations

Collaborative working

| Good at defining problems | Work cooperatively in mutually supportive ways to seek solutions together |

Problems ——————————————— Solutions

| Campaign about issues, state criticisms | Use confrontation to get their solutions accepted |

Campaigning

Some campaigning groups say they now focus on solutions. They can do this in different ways. Deciding the best solutions on their own and then pressing for these solutions to be accepted is easy for those with campaigning skills. Instead of being critical they are being positive but they remain a campaigning organisation. A genuine shift to collaborative problem solving is much harder and requires a different mindset within the voluntary organisation. It also requires very different skills among those who take part in joint problem solving meetings. When a business wants to work with those in the voluntary sector it is worth giving thought to the sort of collaboration that will be most appropriate. Figure 10.2 describes four orientations which can help make the right choice.

LOCAL AGENDA 21

Over two-thirds of local government authorities in the UK are working on their plans for implementing Local Agenda 21 (LA21).[8] An integral part of this planning is the involvement of businesses, voluntary groups and citizens in identifying the priorities, building visions and implementing solutions. There is scope for much more involvement from business. The knowledge and experience of business methods, quite apart from the practical contribution of businesses in different sectors is widely sought.

A typical goal of LA21 is to reduce waste by 25 per cent by the year 2000. If this is not achieved the quantity of waste for disposal will increase. This could mean more lorries taking waste to landfill or the need to build another incinerator. In both cases there is likely to be an increase in road traffic, even more congestion, higher levels of emissions, adverse effects on health and possibly more interference with local trade. If wastes can be reduced this can mean less traffic, less need for landfill sites or an incinerator, and less pollution. The way in which solutions are devised and implemented can help local business. It might be argued that the larger organisations may be the losers and new construction projects could be blocked or become redundant. Against that there is considerable scope to devise ways in which businesses can contribute to the development of new solutions. Local Agenda 21 groups usually welcome more cooperation from businesses.

APPLYING ESI

Figure X

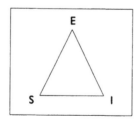

Adopting an ethical approach, linked to enlightened self-interest as the basis of innovation has already been emphasised. Using this framework for taking initiatives can be very productive. For example Electrolux is a good example of a company that has done this. Statements from their Chief Executive and the Vice President of Environmental Affairs are used to illustrate their approach and are summarised in Table 10.2.

Table 10.2 The Electrolux Approach[9]

The Electrolux approach is based on the non-negotiable systems conditions established by The Natural Step described in Chapter 4. These provide strategic guidance for every aspect of their business. Johansson says:

We want to integrate environmental imperatives into management strategy and into the thousands of decisions all our employees make.

Per Grunewald, Senior Vice President Environmental Affairs, adds:

Our environmental strategies have been determined. We are now working intensively to integrate them into our operations.

As a worldwide company Electrolux takes account of different cultures, different stages of maturity and adjusts the pace of change to suit each country.

The Co-operative Bank is another example of successful application of ethical principles for the development of their business (see Chapter 5).

BEYOND LEADERSHIP

Warren Bennis, Jagdish Parikh and Ronnie Lessem have written a stimulating book entitled *Beyond Leadership*.[10] All three authors are well known exponents of progressive practices in management, organisation development and leadership. In their book they put forward ideas that integrate concepts from personal development, organisational learning and sustainable development into a leadership approach. They combine the following ideas:

- self mastery – the competence of the *individual manager;*
- group synergy – *good teamwork* to achieve the best results;
- organisational learning – the ability to *learn as an organisation;*
- societal sustainability – emphasises *sustainable development.*

These four elements are seen as the key to achieving a transition from the old paradigm of conventional business to the new paradigm that tomorrow's society will need. Self-mastery combines intuition and vision to provide the basis for effective action. Good teamwork emphasises the importance of managing situations rather than telling people what to do. It includes sound management of conflict as the key to group synergy. The ability of people in organisations to learn together includes self-mastery and group synergy but adds systems thinking and the ability to concentrate on mastering complexity rather than subordinating people. Sustainable development includes the ability to manage in the age of information and emphasises the importance of ecological, social and economic sustainability. Those who achieve this integration leave the world of traditional leadership and move towards the ideas of beyond leadership. The model used as the basis for *Beyond Leadership* is shown in Figure 10.3.

Figure 10.3 Beyond Leadership

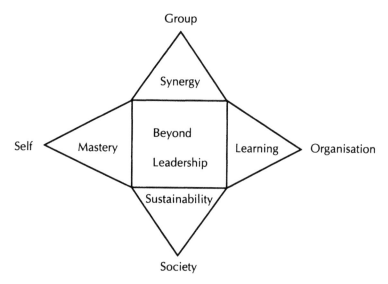

TRANSITIONS

The old ways are changing as more people reach out for a sustainable future. This means that in many areas new thinking and new behaviour is needed. The juxtaposition of traditional thinking and innovative ideas can be a helpful way to appreciate what is involved. For example a summary of the main transitions which have been explored in earlier chapters are set out in Table 10.3.[11]

Table 10.3 Society in Transition

Traditional values and thinking		Emerging values, innovative ideas
Materialism is the key to progress	➤	Materialism alone is not enough
The Earth is resilient	➤	The Earth is threatened
Man can change nature	➤	Natural Laws are immutable
Social trends continue as usual	➤	New social trends are emerging
Profit drives business	➤	Stakeholders want more than profit
Self-centredness prevails	➤	Partnerships are gaining strength
Short-term perspectives	➤	Long-term perspectives

People feel powerless   →   People feel empowered
Learning occurs mainly at school   →   Learning is for life
Adversarial approach to conflict   Creative resolution of conflict

A more comprehensive framework exploring the old and new paradigms is given in Appendix 3.

BUILDING TO LAST

Companies seeking lasting success have known for years that they need to be sound financially. Those wanting a competitive edge know that the competence of their people is equally crucial. Market awareness is a third area which is accepted as vital. However, the inclusive approach, adds a new dimension. It is now important to meet the needs of all stakeholders – shareholders, employees, customers, suppliers and the community. This approach shifts the emphasis of the key customer interface from making transactions to developing relationships. The quality of the relationship with all stakeholders is an essential goal for companies which are building to last.

Alongside the inclusive approach these organisations need to explore the implications of sustainable development. Recognising the ecological imperative and the non-negotiable systems conditions established by The Natural Step is another imperative. The first environmental challenge for companies is to establish environmental policies which lead to cutting waste and emissions. The second challenge is to understand the implications of sustainability. This can only be done effectively when ecological considerations permeate every aspect of the business.

Ecological factors will then be considered alongside financial considerations in every decision. Examples of this have been described earlier in this book and there is little doubt that more will emerge in the years ahead.

The inclusive approach and sustainable development are the twin challenges for future leaders. Both the inclusive approach and sustainable development depend on extensive involvement and understanding of people at every level. This means that organisations will need to create participative and adaptable organisation cultures. The future contains many uncertainties and organisational adaptability will be a crucial factor in building an organisation which lasts.

An example of how a large organisation came to terms with the need for radical change is described by Karl-Henrik Robèrt telling the story about IKEA being introduced to The Natural Step. See Table 10.4.

Table 10.4 IKEA and The Natural Step

IKEA, the Swedish furniture manufacturer and retailer, is an interesting example of how a large, successful business first doubted their ability to adopt The Natural Step's non-negotiable systems conditions, then changed their mind. Karl-Henrik Robèrt, founder of The Natural Step, tells the story.[12]

The management team at IKEA agreed that the systems conditions could be applied to farmers and municipalities but not to IKEA. The reason was that they prided themselves on supplying the cheapest possible furniture to the customer. They were convinced that using recycled metals, decreasing the use of chemicals and plastics, and buying only timber grown in sustainable forests would add greatly to their costs. This would mean that others would undercut IKEA's price and ruin their business. As always Karl-Henrik refused to recommend what they should do because he maintains that those who know the business know more than he does about how to accommodate the conditions for their own unique circumstances.

The breakthrough came when one of the IKEA managers realised that The Natural Step was not being prescriptive but suggesting they go right up to the obstacles and push. He suggested that IKEA could manufacture two ranges of the same product. The original range would not change but they could add a range of products which took account of all four conditions. Both ranges could be the cheapest of their kind. The founder of IKEA wants to deal with this issue in front of the customer by saying:

This is our new sofa. The quality of this new sofa is not only inherent in its construction but extends right across the entire ecosystem. Sure, it's more expensive, but this quality of sofa cannot be found at a better price anywhere else!

IKEA have now resolved never again to ignore the systems conditions and have embarked on an extensive training programme for all their employees. Once this commitment is accepted the practical day-to-day matters start being handled differently. When a garbage contractor failed to clear the rubbish the warehouse manager took the opportunity to empty the container and show his team the contents. Each item could be classified in terms of the systems conditions. Wasteful packaging, metals being thrown away, and hazardous chemicals being taken to landfill were all identified. From that moment the operational strategy was being brought in line with the newly agreed business strategy which is to honour the systems conditions.

LEADERSHIP

Throughout this book examples of outstanding leadership have been given and quotations have been used to illustrate the forward thinking of several chief executives. In particular the inspirational leadership of businessmen with a successful track record in their own spheres is worth noting. They come from many different countries and have made an impressive cintribution towards sustainable development. They include Aurelio Peccei (Italy) cofounder of the Club of Rome, Maurice Strong (Canada) Secretary General of the UN Environmental Conferences in Stockholm (1972) and UNCED in Rio (1992), Stephan Schmidheiny (Switzerland) founder of the BCSD and author of *Changing Course*, Bjorn Stigson (Sweden) Executive Director of the WBCSD and Derek Wanless (UK), former Chairman of ACBE.

It is appropriate to end this chapter and the book by drawing attention to the type of leader that is likely to be successful in the years ahead. Leadership is required in every sector of the community and can come from any level of an organisation.

Three types of situation have been identified in which leadership is required:[13]

- Type 1: The leader can decide using judgement and experience; typically this requires adaptation of existing knowledge and often relates to technical solutions.

- Type 2: The situation can be defined with some clarity but solutions are not clear cut or require strong commitment from other people for successful implementation.
- Type 3: The situation is unclear, often complex and hard to define; sound answers or solutions are even harder to establish and often impact on many people in different ways.

Business examples which relate to these three types of situation are:

- Type 1: Making a decision about a complex, negotiated business deal involving one other party. The central issue is a judgement about the terms of the deal and whether they are acceptable.
- Type 2: Setting up a contract with a supplier who obtains raw materials from various sources. The quality of the solution depends on the supplier understanding what is required and being committed to honouring the terms of the agreement.
- Type 3: In a small town there is one major business which provides much of the local employment but is known to cause severe pollution, with some evidence of health risks to children. The situation is complex, there are uncertainties about the severity of the problems and changes to products and processes are expensive. The evidence about the problems has been accumulating steadily and local people are showing increasing concern. Something needs to be done but neither the definition of the problem nor the solution is clear.

In today's world there are an increasing number of situations, similar to Type 3, which are unclear and hard to define. Appropriate solutions are even more difficult to identify and it appears that the different needs of all parties cannot be satisfied. This unexamined assumption may not always be valid. With appropriate leadership it might be possible to find creative solutions which go a long way to meeting the needs of most of the people with an interest in the outcomes. However, this requires a different kind of leadership initiative from the two previous types of situation. In particular it requires the ability to involve people, listen to them and work with them to agree solutions. This kind of leadership can emerge at any level.

Figure X

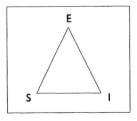

In some Type 3 situations the opportunity to take action rests with the chief executive of the business. There are clear ethical dilemmas but it is also in the self-interest of the business to take action sooner rather than later. A combination of ethics and enlightened self-interest creates the basis for innovative solutions. However, those solutions should meet the different needs of the interested parties who need to be committed to implementing the solutions. The best way to achieve this outcome is to involve people at an early stage in defining the situation and developing possible solutions, then evaluating them and agreeing action steps. This is a classic case for use of the ESI approach.

CONCLUSION

This concluding chapter stresses the importance of taking important initiatives. Businesses have much to contribute to solutions and their initiatives are crucial. Several practical methods are already being used by companies including task forces, awaydays and reviewing their vision and values.

As more organisations identify the need for bringing large groups of people together to develop creative solutions with productive outcomes the methods available for these events are being refined and improved.

In many cases there will be a need to examine the relevance of mindsets for a future that will be different. Changing mindsets is never easy, especially when they seem to have proved successful. However, if they are not appropriate for business in the future it is better to revise them than to continue in the belief that all is well. For the smaller business there are added difficulties of finding the time to do this kind of work but the incentive to get it right is often very strong.

Some companies like the idea of being challenged by people who think differently. This has led to creative, open sessions with campaigners from voluntary organisations of various kinds. Knowing how these people think and work helps the business to choose the right people for effective meetings.

There is great scope for applying the ESI approach and to examine the right leadership approach. Building a business that will last well into the next century requires imaginative management of the transition to the new paradigm. This requires leaders who have the courage to bring people together in uncertain situations with no clear outcomes and jointly work towards agreed solutions.

APPENDIX I: WHY AND HOW SOME
ORGANISATIONS GO FURTHER

Trigger and Organisation	Explanation	Action and Outcome
Values congruency *Merck*[1]	Merck's commitment to the advancement of medical science and service to humanity dates back to 1935 and was restated in 1991.	Mectizan, a cure for 'river blindness' with one million infected people in developing countries, could not be marketed because sufferers and their countries were too poor. It was given away free to all who needed it – no other action would honour Merck's long standing commitment of service to humanity.
Cost structure of business *Rank Xerox*[2]	RX used to lease all their photocopying machines so they remained on their balance sheet. This led to an imbalance between assets and revenue which was getting worse.	Investigation of creative ways to make better use of their assets. To begin with they identified the parts of the machines which could be re-used or recycled. Later they started designing and building machines which could more easily be re-used and recycled. RX are now one of the leading companies in strategic realignment of their business to meet environmental needs.
Impending legislation *Monsanto Chemicals*[3]	In 1986, just prior to tougher USA *legislation, Richard* Mahoney, Chairman of Monsanto pledged to reduce their worldwide air emissions of 320 chemicals by 90 per cent by the end of 1992.	In 1987 Monsanto were emitting 18.4 million pounds of hazardous *chemicals. By 1990 this had been* reduced to 7.8 million pounds and their target was to come down to 1.8 million pounds by the end of 1992. In 1993 Monsanto were able to announce that they had achieved their target and were striving for zero emissions.
Exposed by environmentalists *Norsk Hydro*[4]	Environmentalists climbed over the fence of Norsk Hydro's premises in Norway, took soil samples which they analysed and then	Norsk Hydro were aware of the pollution and had reported it, but it was not in the public domain. In 1990 they published a full report of their emissions and discharges and have gone on to become a leading exponent of environmental

	disclosed their findings to the media. Norway's proudest corporation was exposed as a polluter.	management, with sustainable development firmly on their agenda. They use environmentalists to help train their employees.
Disastrous accident *Union Carbide*[5]	In 1984 methyl isocyanate,poisonous gas escaped from their factory in Bhopal, India, killing at least 2,600 people and causing long-term health problems for 300,000.	Robert Kennedy, CEO of Union Carbide admitted in 1990 that 'people don't trust us'. People generally have a low opinion of the chemical industry and its secrecy. Union Carbide among others have played a leading part in the Responsible Care programme which they use to give details of pollution prevention and their targets for improvement.
Fear of the unknown *B & Q*[6]	B&Q were unable to answer questions from the media on various matters including the source of timber sold in their DIY shops. They had a 'fear of the unknown'.	The reply B&Q gave to media representatives about their timber sources was seen and liked by the Board of B&Q because it contained lots of action. This formed the basis for their environmental policy which has now spread throughout the company and its suppliers. It includes work in the rain forests to help indigenous people gain more from their resources as well as better forest management.
Core business *Body Shop*[7]	From inception Body Shop has had a strong ethical and environmental core.	A business like this needs to maintain standards not only within the business but also with suppliers and those with a retail franchise. Despite periodic attacks Body Shop have persevered, had their work assessed independently and publicised the results.
Disappointment *IBM UK*[8]	IBM worldwide have had a positive approach to environmental matters for many years. In the UK there was disappointment that their carefully prepared environmental reports got little attention.	A new initiative was launched which involved consulting the stakeholder. The results of the survey carried out by ECOTEC were published in September 1995 at a public meeting which ensured some comment and reaction. The survey established benchmarks for the future in 11 areas such as management of the

		environment, product stewardship, energy, transport and IBM's influence on environmental attitudes.
Niche market *BP Solar Energy*	BP have established a Solar Energy Division.	Alongside their mainline oil and chemicals businesses BP have established a successful Solar Energy Division – a good example of a niche market being developed.
Ecological challenge/ insights *Electrolux*[9]	With their Head Office in Sweden Electrolux were well placed to learn from The Natural Step and the non-negotiable system conditions which emerged from the consensus of Swedish scientists (see Chapter 4).	Electrolux were impressed by the ecological challenge as represented by the work of The Natural Step and have undertaken to use the non-negotiable system conditions in the future development of their business. Their environment reports for 1994 and 1995 provide impressive evidence of the work they are doing and, uniquely, describe how their re-focused business strategy enables them to develop profitable business in ways which help the environment.
Selected for renovation *Post Office, Reno, Nevada, USA*[10]	In the mid 1980s Reno Post Office was selected for renovation to make it a 'minimum energy user'. An architectural firm was appointed to do everything possible to reduce energy use.	The building had high ceilings, a black floor and was noisy. Changes included lowering the ceiling, improving the lighting, improving energy saving in other ways and the acoustics were improved. Productivity shot up by 8 per cent, then settled at 6 per cent. The cost of renovation was $300,000. Savings on energy and maintenance came to $100,000 a year and productivity gains amount to about $400,000 a year.
Become lean and clean *Compaq*[11]	Ron Perkins, Facilities Manager, Compaq Computers in the 1980s saved his company about $1 million a year by cutting energy use and pollution – but only after overcoming major obstacles blocking change.	Perkins approached John Gribi, chief financial officer, whose philosophy was 'spend money to save money'. At the time it was unusual for the facilities and other financial people in Compaq to talk to each other. The dialogue managed to change the mindset of 50 per cent return on investment to a new goal of five to seven year payback. Many of the

| | | changes not only saved energy and cut pollution but improved working conditions and lighting. By systematically removing the barriers to productivity massive changes occurred. In 1985 alone productivity improved by 55 per cent with no single cause. |

100% recyclability
Matsushita Electric UK Ltd [12]

Matsushita Electric achieved 100 per cent recylability if its expanded polystyrene packaging materials.

The expanded polystyrene is used to protect television sets and microwave ovens. Landfill sites are reluctant to accept it because it biodegrades very slowly and when incinerated it causes toxic fumes. Duty of Care provisions meant that other options needed to be found. Recycling was achieved with estimated annual cost saving in 1995 of £100,000.

Staff contest
Dow Chemicals [13]

Starting in 1981, Ken Nelson, of Dow Chemicals, Louisiana organised a staff contest to save energy or reduce waste. Initially projects had to cost less than $200,000 and pay for themselves within one year. From 1981 to 1991 1,000 proposals were put forward for peer review.

In the first year 27 approved projects gave a return on investment of 173 per cent, in the second year 32 projects costing $2.2 million returned an average of 340 per cent per year. By eliminating the $200,000 ceiling the scale of projects and the returns increased further. In 1989, 64 projects cost $7.5 million and yielded returns of 470 per cent and even in the twelfth year 143 approved projects gave an average 298 per cent return on investment. The results are attributed to voluntary shop-floor ingenuity rather than management theories and techniques. The sad thing is that after Ken Nelson retired his idea was abandoned – the ideas never spread, not even in Dow Chemicals.

Chief Executive's conscience
Beacon Press

The Beacon Press.

A leader among small firms with an excellent environmental record.

Closing the loop *Nissan Motor Manufacturing (UK) Limited*[14]	Nissan Motors adopted a 'closed loop' and recycling system for plastic off-cuts.	In 1993 they saved in excess of £700,000 on raw materials, recovered the capital cost within 15 months.
New business opportunity *Reclamation Services Ltd*[15]	Identified a new business opportunity to recover and sell building material to a worldwide market.	This is an outstanding success story of architectural salvage. York flagstones, structured hardwoods, oak block flooring, hardwood beams and trusses, garden statuary and so on are marketed worldwide. The firm also undertakes controlled dismantling of old buildings and has a separate division which designs, produces, repairs and restores ornamental stone pieces.
Avoiding over exploitation *Unilever*[16]	In February 1996 Unilever, the Anglo-Dutch consumer goods group with about 20 per cent of the European and US frozen fish market and global sales worth £600 million, agreed to work with WWF to establish the Marine Stewardship Council to strive for sustainable fish catches.	The Marine Stewardship Council (MSC) will work towards the use of a logo on retail goods to indicate that the fish being sold come from accredited fishing grounds. The overall aim is to ensure the long-term viability of global fish populations and the health of the marine ecosystems on which they depend. The MSC will be set up along the same lines as the Forest Stewardship Council and will aim to arrest the catastrophic decline in fish catches all around the world. Unilever has pledged to source their fishery products only from sustainable well-managed fisheries certified to MSC standards by the year 2005.
Core objective *The Royal Academy of Engineering*[17]	One of the objectives of The Royal Academy of Engineering is 'to promote the advancement of engineering for the benefit of society.' It is appropriate that they should be 'at the forefront of the debate which will crystallise engineering's agenda for	In June 1995 The Council of Academies of Engineering and Technological Sciences (CAETS) published a Declaration on The Role of Technology in Environmentally Sustainable Development.[18] In September 1995 The Royal Academy of Engineering organised a conference on Engineering for Sustainable Development. The conference included papers from a wide cross

	delivering Sustainable Development.'	section of interested parties and was divided into four parts: The Expectations of Society, The Challenge for Manufacturing, The Challenge for Transport and the Issues for Engineers. The Chairman in his conclusion said 'Engineers will turn the vision of Sustainable Development into reality in many, many ways.'
Market opportunity – services *National Provident Institute* [19]	NPI, founded in 1835 is a mutual life office with a progressive approach to pensions and investment business, a network of branches across the UK and over £7 billion invested for 500,000 policyholders. They launched their Global Care family of funds to expand their ethical portfolio.	They attracted a highly regarded research team, led by Tessa Tennant, to join NPI in 1995. They evaluate company records to determine which could be included in the ethical portfolio. The assessment covers financial performance, ethical considerations and environmental impact as well as improvement plans in all three areas. NPI's Global Care Fund is assessed by Holden Meehan in the Independent Guide to Ethical and Green Investment Funds with a top rating in terms of ethics, environment, financial pedigree and the resources applied to screening – the only fund to get a top rating in all four areas. [20]
Waste packaging *Procter & Gamble*	Having achieved a lot with their own business using the slogan 'more from less' P&G turned their attention to what happens to their (and others) waste packages.	P&G became the leading firm working with Adur District Council on a scheme to recover household dustbin waste. The government target is to recycle 25 per cent of 1990 levels by the year 2000. The Adur scheme was a success and much was learned. A test market was then carried out in Worthing. The government target is not achieved by recycling packaging and newspapers alone but is much cheaper than expected. [21]
Crisis *Smith & Nephew* [22]	The Women's Environment Network attacked Smith & Nephew over involvement with	Smith & Nephew challenged the facts on which the attack was based but the programme was broadcast. The adverse publicity led to further reviews within Smith & Nephew on

	dioxins and this attack was taken up by World in Action on TV.	environmental matters. These reviews led to wide-ranging effective environmental policies and their implementation, closely linked to business strategy.
Underestimating the customer *Shell UK* and Brent Spar [23]	When the Brent Spar, a structure used by Shell for extracting oil from the North Sea, was no longer required they obtained scientific advice, approved by the UK government, to dump it at sea. Public opinion differed and Shell products were boycotted – especially in Germany.	Shell reversed their decision to sink the Brent Spar in the sea, damaging their reputation and embarrassing the UK government. The UK government also lost credibility by their support for Shell's original plan. The incident led Shell to publicly apologise to their customers and to the government. They set up an elaborate process with independent mediators to find the best solution for disposal and sought ideas from far and wide. The mediation process is still continuing as this is being written. Greenpeace, who challenged Shell, also made an error and had to apologise.

APPENDIX 2: SMALL FIRMS
AND THE ENVIRONMENT

The contribution of small firms to the UK economy is very significant and the trends suggest that this significance is increasing as the facts, drawn from the Groundwork Status Report on Small Firms and the Environment, show:[1]

- In 1993 there were 3.6 million businesses in the UK – 97 per cent with less than 20 employees.
- Only 3200 businesses employed more than 500 people.
- The DTI estimates for 1993 show that:
  - 99.5 per cent were MICRO or SMALL firms;
  - 0.5% per cent MEDIUM or LARGE firms.
- Turnover of businesses in the UK in 1992/3 based on VAT returns:
  - Under £1,000,000 turnover   92 per cent
  - Over £1,000,000 turnover   8 per cent
- In the UK there are 3100 small firms per million inhabitants – less than in Germany, Japan and the USA.
- Between 1985 and 1989 firms employing less than 20 people created over 1 million jobs in the UK.
- In 1994, 446,000 small firms started up, 420,000 closed down – a net increase of 26,000.
- Seventy per cent of SMEs are 'fairly' or 'very' concerned about their companies' environmental performance.
- Individually SMEs have little impact on the environment but collectively the impact is significant.
- Most environmental pressure is exerted on SMEs by regulations, customers, employees and investors.
- Twenty-seven per cent of SMEs are unaware of environmental legislation, 27 per cent are aware of the Environmental Protection Act. Fewer are aware of Chemical Hazard & Packaging Regulation, local authority air pollution control and Duty of Care Regulations.
- SMEs are most likely to be motivated to adopt environmental policies by:
  - cost savings;
  - legislation and regulatory pressure;
  - directors' liability;
  - market opportunities;
  - gaining a positive company image; and
  - customer pressure.
- SMEs perception of actual and potential benefits from pursuing environmental action are:
  - helping the environment;
  - improved image;
  - higher employee morale;
  - cost savings;

- – compliance with legislation; and
- – increased sales.
- Advice and practical assistance most useful to SMEs is:
  - – company specific advice and improved technology;
  - – personal visit by local adviser;
  - – step-by-step guide; and
  - – discussion with regulator, telephone helpline and printed information.

APPENDIX 3: MOVING TO
THE NEW PARADIGM

Building to last will mean leaving the old paradigm and moving to a new paradigm. This is likely to apply to organisations of any size in the public, private and voluntary sectors. The contrast between the old and new ways of thinking indicates what this could mean in practice. This does not imply any instant transition to the new paradigm, nor that every element is appropriate to every organisation. The framework is a starting point for developing plans to suit particular organisations and the unique situations they face. The framework can be marked to show the current position with a square and the desired position with a circle.

Old thinking	Transition stages	New thinking
World View		
Fragmented world view	1 2 3 4 5 6 7	Holistic world view
Simplistic thinking	1 2 3 4 5 6 7	Complex, systems thinking
Values implicit	1 2 3 4 5 6 7	Values stated explicitly
Aid and care for the poor	1 2 3 4 5 6 7	Equity essential for stability
Business Purpose and Strategy		
Reactive and short-term	1 2 3 4 5 6 7	Proactive and far-sighted
Mission assumed or implied	1 2 3 4 5 6 7	Mission reviewed and restated
Trade and profits take precedence	1 2 3 4 5 6 7	Health and ecology take precedence
Traditional and wasteful	1 2 3 4 5 6 7	Innovative and efficient
Agenda set by competitors or crises	1 2 3 4 5 6 7	Agenda set by foresight
Secretive	1 2 3 4 5 6 7	Open
Goods move long distances	1 2 3 4 5 6 7	Goods move short distances
Selfish self-interest	1 2 3 4 5 6 7	Enlightened self-interest
Hierarchical leadership	1 2 3 4 5 6 7	Participative leadership
Legitimacy unquestioned	1 2 3 4 5 6 7	Licence to operate
Accountability to shareholders	1 2 3 4 5 6 7	Accountability to stakeholders
Monetary measurement of performance	1 2 3 4 5 6 7	Multiple performance measures
Ecological Sanity – Environmental Sustainability		
Increasing dependence on fossil fuels	1 2 3 4 5 6 7	Switching to renewable energy
Increasing dependence on minerals and mined resources	1 2 3 4 5 6 7	Reducing dependence on minerals and mined resources
Increasing dependence on unnatural goods and toxic compounds (non-biodegradable)	1 2 3 4 5 6 7	Increasing switch to natural, non-toxic goods (biodegradable)
Environmental biodiversity diminished	1 2 3 4 5 6 7	Biodiversity encouraged
High level of emissions and wastes	1 2 3 4 5 6 7	Zero wastes and emissions target
Wastes sent to landfill	1 2 3 4 5 6 7	Wastes avoided, re-used and recycled

Social Responsibility and People-Centred Development

Dominating, top down direction	1 2 3 4 5 6 7	Participative management
Adversarial approach	1 2 3 4 5 6 7	Reciprocal relationships
People are directed	1 2 3 4 5 6 7	People are involved
Dependency	1 2 3 4 5 6 7	Empowerment
Top down development	1 2 3 4 5 6 7	People-centred development
Learning predominantly at school	1 2 3 4 5 6 7	Lifelong learning for all

Economic Viability

Economic globalisation	1 2 3 4 5 6 7	Local self-reliance
Market economics	1 2 3 4 5 6 7	New economics
Growth is essential, often indiscriminate	1 2 3 4 5 6 7	Selective growth
External costs ignored	1 2 3 4 5 6 7	All costs internalised
Over reliance on GDP	1 2 3 4 5 6 7	GDP replaced by ISEW

NOTES

NOTES TO INTRODUCTION

1 James C Collins and Jerry I Porras, *Built to Last*, Century, London, 1994.

2 John P Kotter and James L Heskett, *Corporate Culture and Performance*, The Free Press, New York 1992.

3 Gary Hamel and CK Prahalad, 'Competing for the Future', *Harvard Business Review*, July/August, 1994. A book of this title is also available.

4 See for example Peter Drucker, *The Concept of the Corporation*, Mentor Executive Library, New York, 1946; *The Practice of Management*, Mercury Books, London, 1955; and *Managing for Results*, Heinemann, London, 1964. See also Tom Peters and Robert Waterman, *In Search of Excellence*, Harper and Row, New York, 1982.

5 See for example 'Polluting Industries Marked by Index', *The Financial Times*, 18 November 1996; the International Chamber of Commerce list of companies that have signed the Business Charter for Sustainable Development.

6 See for example David Korten, *When Corporations Rule the World*, Earthscan, 1995; and Andrew Rowell, *Green Backlash*, Routledge, 1996.

NOTES TO CHAPTER I

1 Max Nicholson, *The New Environmental Age*, Cambridge University Press, Cambridge, UK; New York and Melbourne, 1987 and John McCormick, *The Global Environmental Movement*, Belhaven Press, London, 1989, provide much of the information in this section.

2 Elspeth Huxley, *Peter Scott*, Faber and Faber, London and Boston, 1993.

3 Rachel Carson, *Silent Spring*, Houghton Mifflin, 1962 and Penguin,1965.

4 Committee on Resources and Man, *Resources and Man*, National Academy of Sciences − National Research Council, W H Freeman and Co, San Francisco, 1969, p8.

5 Report of the Study of Critical Environmental Problems, *Man's Impact on the Global Environment: Assessment and Recommendations for Action*, Massachusetts Institute of Technology, 1970.

6 Most of the reports associated with the Club of Rome were commissioned by them. However, the Council of the Club of Rome produced their own report summarising the global 'problematique'. This was written by Alexander King and Bertrand Schneider, entitled, *The First Global Revolution*, published by Simon and Schuster, 1991.

7   Donnella Meadows, Dennis Meadows, Jorgen Randers and William Behrens III, *Limits to Growth*, Earth Island Ltd, London, 1972.

8   Barbara Ward and Rene Dubos, *Only One Earth*, Penguin Books, Harmondsworth, 1972.

9   Edward Goldsmith, Robert Allen, John Davoll and Sam Lawrence, 'The Blueprint for Survival', published as the January 1972 edition of *The Ecologist.*

10 E F Schumacher, *Small is Beautiful*, published by Blond and Briggs,1973.

11 Information about the 'Schumacher Circle' is available from the Schumacher Society, Ford House, Hartland, Bideford, Devon EX39 6EE, UK.

12 Amory Lovins and Hunter Lovins, *World Energy Strategy*, 1975, *Soft Energy Paths*, 1977 – both now out of print. However, more recent material is available from Rocky Mountain Institute, 1739 Snowmass Creek Road, Snowmass, Colorado 81654–9199, USA.

13 Lester Brown et al, *Vital Signs 1996 –1997*, Earthscan, London, 1996, p54.

14 Barbara Ward, *The Home of Man*, Penguin Books, Harmondsworth, 1976.

15 Council on Environmental Quality, *The Global 2000 Report to the President: Entering the Twenty-first Century*, Harmondsworth, Penguin, 1982.

16 Willy Brandt, Chairman of the Independent Commission on International Development Issues, *North-South: A Programme for Survival*, Pan, London, 1980, and *Common Crisis: Cooperation for World Resources*, Pan, London, 1983.

17 IUCN/UNEP/WWF, *World Conservation Strategy: Living Resource Conservation for Sustainable Development*, published by IUCN/UNEP/WWF, Gland, Switzerland, 1980. Summarised by John McCormick, *The Global Environmental Movement*, Belhaven Press, London, 1989, pps167–169.

18 Lester Brown et al, *State of the World* and *Vital Signs*, published annually by W W Norton in the USA and Earthscan in the UK. The data from the Worldwatch Institute publications is available on disk – enquiries to the book publishers.

19 World Commission on Environment and Development, *Our Common Future*, Oxford University Press, 1987. This book is also known as the Brundtland Report.

20 John Elkington and Julia Hailes, *The Green Consumer Guide*, 1988 and *The Green Consumer's Supermarket Shopping Guide*, 1989, published by Victor Gollancz, London.

21 Phil Wells and Mandy Jetter, *The Global Consumer: best buys to help the Third World*, Victor Gollancz, 1991.

22 OECD, *The State of the Environment*, OECD, Paris, 1991.

23 IUCN/UNEP/WWF, *Caring for the Earth: A Strategy for Sustainable Living*, IUCN/UNEP/WWF, Gland, Switzerland, 1991.

24 Stephan Schmidheiny, *Changing Course: A Global Business Perspective on Development and the Environment*, MIT, Cambridge, Massachusetts and London, UK, 1992, ppsxx–xxi.

25 Michael Grubb, Matthew Koch, Abby Munson, Francis Sullivan and Koy Thomson, *The Earth Summit Agreements: A Guide and Assessment*, published by Earthscan for The Royal Institute of International Affairs, London, 1993.

26 Donella Meadows, Dennis Meadows and Jorgen Randers, *Beyond the Limits: Global Collapse or a Sustainable Future*, Earthscan, London, 1992.

27 Al Gore, *Earth in the Balance: Forging a New Common Purpose*, Earthscan, London, 1992, p366.

28 Deborah MacKenzie, 'Stormy Weather for Insurers', Tomorrow, April – June 1993; p32.

29 John Houghton, *Global Warming: The Complete Briefing*, Lion Books, Oxford, 1994, p17; and Al Gore, *Earth in the Balance*, Earthscan, London, 1992, Chapter 15.

30 Available from the International Chamber of Commerce, 38 Cours Albert 1er, 75008 Paris, France, also from national offices of the ICC.

31 Jan-Olaf Willums and Ulrich Goluke, *From Ideas to Action*, International Chamber of Commerce, 1992.

32 Tomorrow Publishing, Halsingegatan 9, SE–113 23 Stockholm, Sweden. They have regional offices in France, Switzerland and Singapore.

33 In addition to the ICC book by Jan-Olaf Willums see also: Julie Hill, Ingrid Marshall and Catherine Priddey, *Benefiting Business and the Environment*, published by the Institute of Business Ethics, 12 Palace Street, London SW1E 5JA, 1994.

34 Advisory Committee on Business and the Environment, *Sixth Progress Report to and Response from the President of the Board of Trade and the Secretary of State for the Environment*, published by the Department of Trade and Industry, London, UK.

35 Institute of Management Sounding Board, *Management Today*, September 1996, p5.

36 Business in the Community and Business in the Environment, 8 Stratton Street, London W1X 5FD, UK.

37 *Tomorrow's Company Report* (1995) is available from RSA, 8 John Adam Street, Adelphi, London WC2N 6EZ.

38 *Business and the Environment: In the Aftermath of Brent Spar and BSE*, lecture given by Robert Worcester, Chairman of MORI, to HRH Prince of Wales's Business & the Environment Programme, University of Cambridge, September, 1996.

39 Peter Martin and Colin Hutchinson, *Towards Sustainability*, Office for Public Management, 252b Gray's Inn Road, London WC1X 8JT.

40 Francis Kinsman, *The New Agenda*, Spencer Stuart & Associates, 1983. This is described at a meeting where the author knew from a series of individual meetings that those present held similar views to each other. Despite assurances of confidentiality none was prepared to express their views at the meeting.

41 See John Kotter and James Heskett, *Corporate Culture and Performance*, Free Press, 1992. My own experience of working with clients on culture change is described in *Vitality and Renewal: A Manager's Guide for the 21st Century*, Adamantine Press UK and Praeger, USA, 1995, Chs 9 to 11.

42 Ann Goodman, 'King Salomon Minds', *Tomorrow*, no 5 vol VI, September – October 1996, p46.

43 Lester Brown et al, *Vital Signs 1996/97*, Earthscan, London and W W Norton, USA, provide much of the information in this section. Other sources include Lester Brown et al, *State of the World 1996*, and John Houghton, *Global Warming*, Lion Books, 1994.

44 John Houghton, *Global Warming: The Complete Briefing*, Lion Books, Oxford, 1994, pp31 and 46.

45 Lester Brown, *State of the World 1994*, Earthscan, London, Ch 10 and p191. See also Lester Brown, *Who Will Feed China?* Earthscan, London, 1995.

46 Lester Brown, *State of the World 1996*, Earthscan, London, 1996, p7.

47 For example, 'First steps towards a green GDP', *Financial Times*, 30 August 1996, and the work of the New Economics Foundation.

48 Al Gore, *Earth in the Balance*, Earthscan, London, 1992, pp269, 270.

NOTES TO CHAPTER 2

1 The New Economics Foundation, First Floor, Vine Court, 112–116 Whitechapel Road, London E1 1JE. Tel: 0171 377 5696.

2 See for example David Pepper, *Modern Environmentalism*, Routledge, London and New York, 1996. Appendix 3 contains a more detailed analysis of a paradigm shift.

3 Holden Meehan, *An Independent Guide to Ethical and Green Investment Funds, sixth edition*, 1996, Holden Meehan, 11th Floor, Clifton Heights, Triangle West, Clifton, Bristol BS8 1BR. Tel: 0117 925 2874 or 0171 242 0226.

4 Figures provided by Ethical Investment Research Service (EIRIS) and rounded to give six month total investment figures. See also Russell Sparkes, *The Ethical Investor*, Harper Collins, London, 1995.

5 *Money & Ethics* 1996 edition, by Peter Webster, published by EIRIS, 504 Bondway Business Centre, Bondway, London SW8 1SQ.

6 'Green Funds: Fuzzy No More', *Tomorrow*, no 5, vol VI, September/October 1996, p24.

7 'Reaping the Profits of Tomorrow's World', *The Observer*, 24 November 1996.

8 Lester Brown et al, *Vital Signs 1966/1967*, Earthscan, London, 1996 pp48–61.

9 Figures from Global Action Plan UK, Energy Action Pack.

10 Lester Brown, et al, *State of the World 1996*, Earthscan, London, 1996, pp157–161.

11 Michael Jacobs, *The Politics of the Real World*, published by Earthscan, London, 1996, pp96–100 and 'Witness Box 18', p99.

12 Ed Mayo, *Community Banking: A Review of the International Policy and Practice of Social Lending*, New Economics Foundation, First Floor, Vine Court, 112–116 Whitechapel Road, London E1 1JE; undated.

13 Dauncey, Guy, *After the Crash*, Greenprint, London, 1988, p52–69.

14 Letslink UK, 61 Woodcock Road, Warminster, Wiltshire BA12 9DH. Tel: 01985 217871

15 David Pepper, *Modern Environmentalism*, gives a wider review, see note 2.

16 GAP International, Stjarnvagen 2, S–182 46 Enebyberg, Sweden. Global Action Plan, Inc, P O Box 428, Woodstock, NY 12498, USA.

17 Global Action Plan UK, 8 Fulwood Place, London WC1V 6HG. Tel: 0171 405 5633.

18 *Which?* Magazine, August 1996, p12–17.

19 Vegetarian Society annual report 1995.

20 Statistics provided by The Vegetarian Society, Parkdale, Dunham Road, Altrincham, Cheshire WA14 4QG.

21 Lester Brown et al, *State of the World 1994*, Earthscan, London,1994 p192.

22 *Which?* survey, November 1995 and report in *Observer*, 18 February 1996.

23 The Soil Association, Organic Food and Farming Centre, 86–88 Colston Street, Bristol BS1 5BB. Tel: 0117 929 0661.

24 Lester Brown et al, *Vital Signs 1996–1997*, published by Earthscan, London, 1996, p110.

25 Nic Lampkin, University of Wales, Aberystwyth, Aberystwyth SY23 3AL.

26 Henry Doubleday Research Association, Ryton Gardens, Ryton-on-Dunsmore, Coventry CV8 3LG. Tel: 01203 303517.

27 Sustainable Agriculture, Food and Environment Alliance, 38 Ebury Street, London SW1W 0LU. Tel: 0171 823 5660.

28 *The Politics of the Real World*, by Michael Jacobs on behalf of The Real World Coalition, published by Earthscan, London, 1996.

29 Jack Schofield, *Observer: The Business*, Sunday 5 May, 1996 is just one source of information and predictions.

30 British Telecom leaflet, 1991.

31 This information comes from conversations with Charter 88 and their publication *Citizens.*

NOTES TO CHAPTER 3

1 The Green Alliance, *The 1996 UK Business & the Environment Trends Survey,* sponsored by Entec and conducted by The Moffatt Associates Partnership, 72 Boston Place, London NW1 6EX.

2 Available from the RSA, 8 John Adam Street, London WC2N 6EZ.

3 Shann Turnbull *New Economics,* Issue 39, Autumn 1996, drawing on the work of Professor Max Clarkson, University of Toronto.

4 Contact through the RSA until Tomorrow's Company Centre is established – see note 2.

5 Robert Worcester, *Business and the Environment: In the Aftermath of Brent Spar and BSE,* HRH The Prince of Wales's Business & the Environment Programme, University of Cambridge, 16 September 1996.

6 Simon Bryceson, '*Football between the Trenches*', Tomorrow, no 3 vol VI, May/June 1996, p30.

7 Francis Kinsman, *Millennium,* W H Allen, Chatham, 1990.

8 Follow-up surveys in Europe are now conducted by Synergy Consulting, 54–58 Uxbridge Road, Ealing, London W5 2TL. They expect to publish results of a new survey in 1997.

9 Synergy Consulting – see note 8.

10 Based on Synergy Consulting findings – see note 8.

11 See for example Julie Hill, Ingrid Marshall and Catherine Priddey, *Benefiting Business and the Environment,* Institute of Business Ethics, 1994.

12 See note 11 above *Benefiting Business and the Environment.*

13 John Elkington, *Who Needs It?,* SustainAbility, 1995.

14 A particularly stimulating book is *Imaginization: the art of creative management* by Gareth Morgan, Sage Publications, USA, UK and India, 1993.

15 Dorothy Mackenzie, *Green Design,* Laurence King, London, 1991. A new edition is expected in 1997.

16 Victor Papanek, *The Green Imperative: Ecology and Ethics in Design and Architecture,* Thames and Hudson, London, 1995.

17 Dorothy Mackenzie, *Green Design*, op cit, Case Study pp56–59.

18 Lester Brown et al, *State of the World 1995*, Earthscan, London, Ch 6 & p96.

19 Based on Peter Martin's model described in *Towards Sustainability*, by Peter Martin and Colin Hutchinson, Office for Public Management, 1996.

20 Examples include the work of The World Business Council for Sustainable Development, Claude Fussler and Peter James, *Driving Eco-management*, Pitman, London, 1995; and Procter & Gamble's 1995 report, *Design Waste Out*.

21 Papanek, *The Green Imperative*, Thames and Hudson, London, 1995, p17.

Notes to Chapter 4

1 This chapter draws on the Worldwatch Institute's annual publications *State of the World* and *Vital Signs, Caring for the Earth* by UNEP, WWF and IUCN (1991) and *The Politics of the Real World* by The Real World Coalition (1996), all published by Earthscan, London. The application of sustainable development by business focuses attention on building to last.

2 'Sustainable Corporations? Isn't that an oxymoron?', by Stephen Viederman, *Tomorrow* No 4, vol V, July/August 1996.

3 Published by HMSO.

4 *Sixth Progress Report to and Response from the President of the Board of Trade and the Secretary of State for the Environment*, HMSO, April 1996.

5 *Indicators of Sustainable Development for the United Kingdom*, Department of the Environment and Government Statistical Office, HMSO, March 1996.

6 World Commission on Environment and Development, *Our Common Future*, Oxford University Press, 1988.

7 UK Round Table on Sustainable Development, *First Annual Report*, April 1996.

8 Colin Hutchinson, *Vitality and Renewal: A Manager's guide for the Twenty-first Century*, Adamantine (UK) and Praeger (USA), 1995, p101.

9 Ervin Laszlo, *The Inner Limits of Mankind*, Oneworld, 1989, p25. This is an excellent short book which challenges conventional thinking and provides pointers to the future.

10 Ehrlich, P and Holdren J, 1971, 'The Impact of Population Growth,' *Science*, vol 171 pp1212–17. It is further discussed in Ehrlich, P and Ehrlich A, *The Population Explosion*, 1990, published by Hutchinson, pp58–59 and subsequent pages and Ehrlich and Ehrlich, *Healing the Planet*, 1991, published by Addison-Wesley, pp6 and subsequent pages.

11 Michael Jacobs, *Politics of the Real World*, published by Earthscan, London, 1996, p27.

12 This figure might seem absurdly high and completely unattainable but Monsanto, the giant chemical company set a goal to cut their waste by 90 per cent in 1986 and achieved it by 1992 – see Jan-Olaf Willums, *From Ideas to Action*, International Chamber of Commerce, published by Earthscan, London, 1995, p116.

13 For a fuller discussion see *Caring for the Earth* published in partnership by IUCN – The World Conservation Union, UNEP – United Nations Environment Programme and WWF – World Wide Fund for Nature, 1991. See also *Beyond the Limits* by Donella Meadows, Dennis Meadows and Jorgen Randers, published by Earthscan, London, 1992.

14 Mathis Wackernagel and William Rees, *Our Ecological Footprint*, published by New Society Press, Gabriola Island, BC and Philadelphia, PA, 1996; or in the UK from Jon Carpenter Publishing, The Spendlove Centre, Charlbury OX7 3PQ. Tel/Fax: 01608 811969.

15 Donnella Meadows, Dennis Meadows and Jorgen Randers, *Beyond the Limits*, Earthscan, London, 1992, p123 and p210.

16 This is vividly illustrated by the winners and losers from the GATT Uruguay Round and the projected gains and losses from trade liberalisation to the year 2000. See Jacobs, *The Politics of the Real World*, Earthscan, London, 1996, p49. This shows benefits to the EU of US$80 billion compared with less than US10 billion for Latin America, India and Eastern Europe and a loss of some US$3 billion in Africa.

17 For a fuller discussion see Holmberg, J; Robert, K-H and Eriksson K-E, *Journal of Ecological Economics*, Chalmers Technical Institute. See also The Natural Step, *A Collection of Articles*, published by The Natural Step, available from Der Naturliga Steget, Amiralitetshuset, Skeppsholmen, S–111 49 Stockholm, Sweden or from The Natural Step for Britain, Forum for the Future, Thornbury House, 18 High Street, Cheltenham, GL50 1DZ.

18 Several practical examples are provided by Walter Wehrmeyer (editor), *Greening People*, Greenleaf Publishing, Sheffield, 1996.

19 Carl Frankel, 'A Sustainable SEA Change', *Tomorrow* no 4, vol VI, July/August 1996.

20 Supercars: The Coming Light-vehicle Revolution, presented to the European Council for an Energy Efficient Economy, Rungstedgard, Denmark, June 1993

21 Rocky Mountain Institute Newsletter, Spring 1996.

22 Adapted from Willis Harman, 'A System in Decline or Transformation', *Perspectives* 8(2), World Business Academy, 1994.

NOTES TO CHAPTER 5

1 Colin Hutchinson, 'Integrating Environment Policy with Business Strategy', *Long Range Planning*, vol 29 no 1 pp11–23, February 1996. This article stimulated the

idea for this book and this chapter, in particular, builds on the ideas that it contained.

2 Jan-Olaf Willums and Ulrich Goluke, *From Ideas to Action*, ICC, 1992, pp116 and 186.

3 Assessments of Environmental Reports have been carried out by *Tomorrow* magazine, vol V, no 1, vol V, no 3, no 4 and no 6, also *UK Environmental Reporting Survey*, KPMG, 8 Salisbury Square, London EC4Y 8BB.

4 John Elkington and Shelly Fennell, 'Verification: can credibility be bought?' *Tomorrow* vol VI no 5, September–October 1996.

5 World Business Council for Sustainable Development supplement contained in *Tomorrow* vol V no 3, July–September 1995. WBCSD has a supplement in every issue of *Tomorrow* – see note 8 for contact address.

6 Advisory Council for Business and the Environment, Fourth and Fifth Reports, October 1994 and July 1995 offer several recommendations. The reports are available from the Department of the Environment, London.

7 'Getting the Financial Markets to Support Sustainable Development', *Tomorrow* vol V no 3 , WBCSD Supplement.

8 *Tomorrow*, Tomorrow Publishing, Hasingegatan 9, SE–113 23 Stockholm, Sweden.

9 The 5Cs formulation is contained in *The Environment at Work: Caring for the Earth is Your Business*, by Colin Hutchinson, published by Chartwell-Bratt Ltd, Old Orchard, Bickley Road, Bromley, Kent BR1 2NE. This is a one day awareness-to-action programme for any company and has been adapted for use in the public sector.

10 Chris Marsden, 'Competitiveness and Corporate Social Responsibility', published in *Organisations and People* vol 3 no 2, May 1996.

11 Quoted in Stephan Schmidheiny, *Changing Course*, MIT, 1992, p48.

12 These figures have been confirmed by the named companies.

13 Collins and Porras, *Built to Last*, Century, London, 1994, p1.

14 P&G 1995 Environmental Progress Report.

15 Peter White, *Sustainable Product Lifecycles, Engineering for Sustainable Development*, edited by Dr James McQuaid, The Royal Academy of Engineers, 1995. See also 'Procter & Gamble: Environmental and Social Responsibility' and 'Environmental management in an international consumer goods company' by P R White, B De Smet, J W Owens and P Hindle published in Resources, Conservation and Recycling, no 14, 1995.

16 *Rank Xerox Fact Book*, Rank Xerox Ltd, Parkway, Marlow, SL7 1YL.

17 Rank Xerox, The Document Company, Environmental Performance Report, November 1995.

18 *Electrolux and the Environment 1994: Vision, Policy and Steps Taken*, Electrolux, Lilla Essingen, S–105 45 Stockholm, Sweden.

19 Electrolux Environmental Annual Report 1995, Lilla Essingen, S–105 45 Stockholm, Sweden.

20 The Co-operative bank, *Environmental Services for Business Customers*, National Centre for Business and the Environment, 1 Balloon Street, Manchester M60 4EP, UK.

21 *Purpose beyond Profit*, The Co-operative Bank, 1 Balloon Street, Manchester M60 4EP.

22 *How Green is My Front Door? B&Q'sSecond Environmental Review*, B& Q plc, Portswood House, 1 Hampshire Corporate Park, Chandlers Ford, Eastleigh, Hampshire SO53 3YX, published July 1995. This report describes how B&Q have applied their environment policies to timber, peat, product and packaging, retail operations, the international dimension and making the environment mainstream. Selective use of this data informs the B&Q section of this chapter.

23 Kim Loughran interview, 'The Green Knight', *Tomorrow* vol 5 no 4 , October–December 1995.

24 B&Q publication, *From Growing Tree to Point of Sale*, and Willums and Goluke, *From Ideas to Action* – see note 2.

NOTES TO CHAPTER 6

1 Stephan Schmidheiny, *Changing Course*, MIT, 1992, p221.

2 Schmidheiny, *Changing Course*, p221/2.

3 Schmidheiny, *Changing Course*, p221.

4 *Chemicals in A Sustainable World*, Chemical Industry Association, King's Buildings, Smith Square, London SW1P 3JJ, 1993.

5 Bruce Ames in an interview with Richard North, *A Green with Spleen*, The Independent 17 July 1996.

6 Paul Hawken, *The Ecology of Commerce*, Weidenfeld & Nicolson, London, 1993, p55.

7 Norman Myers, General Editor, *Gaia Atlas of Planet Management*, Gaia Books Limited, 1994, p 120/1.

8 Geoffrey Lean and Don Hinrichsen, *Atlas of the Environment*, Helicon, 1992, p101.

9 Geoffrey Lean, *Deadly Peril in our Culture of Denial*, Independent on Sunday 24th March, 1996.

10 BBC Wildlife Magazine, March 1996. See also *Bad Harvest?* By Nigel Dudley, Jean-Paul Jeanrenaud and Francis Sullivan, Earthscan, London,1995.

11 Francis Sullivan, 'Beyond 1995', leader of WWF's Global Forest for Life Campaign, BBC Wildlife March 1996, p77.

12 WBCSD, 160, route de Florissant, C H-1231 Conches–Geneva, Switzerland. Both the summary document and the main report are available from E&Y Direct, Unit 5, The Alpha Centre, Upton Road, Poole, Dorset, BH17 1XR.

13 WWF's Endangered Seas Campaign, WWF International, Panda House, Weyside Park, Godalming, Surrey GU7 1XR, UK.

14 Lester Brown et al, *Vital Signs 1996–1997*, Earthscan, London, 1996, p30–33.

15 See Worldwatch Institute's *State of the World* and *Vital Signs*, both published annually by Earthscan.

16 *Financial Times*, 'Unilever in fight to save global fisheries', 22 February 1996.

17 Michael Sutton and Caroline Whitfield, *The Marine Stewardship Council: New Hope for Marine Fisheries*, from Michael Sutton, 1996, WWF, Panda House, Weyside Park, Godalming, Surrey GU7 1XR, UK, or Caroline Whitfield, C/0 Birds Eye Wall's Ltd, Station Avenue, Walton-on-Thames, Surrey KT12 1NT.

18 The International Hotels Environment Initiative, 5 Cleveland Place, St James, London SW1Y 6JJ, UK.

19 *The Role of Technology in Environmentally Sustainable Development*, a Declaration of the Council of Academies of Engineering and Technological Sciences (CAETS), available from The Royal Academy of Engineering, 29 Great Peter Street, Westminster, London SW1P 3LW. The other signatories are the comparable Academies in Australia, Belgium, Canada, Denmark, Finland, France, Japan, Mexico, The Netherlands, Norway, Sweden, Switzerland and The United States of America.

20 Dr James McQuaid (ed), *Engineering for Sustainable Development*, The Royal Academy of Engineering, 1995: 29 Great Peter Street, London SW1P 3LW.

21 Building Research Establishment, Garston, Watford, WD2 7JR, UK.

22 See note 21– same address as BRE.

23 News of Construction Research, April 1996, the BRE newsletter.

24 *The Times 1000*, Times Books, London, 1996.

25 *Engineering for Sustainable Development*, edited by Dr James McQuaid, p91, see note 20.

26 The information for this section comes from a telephone conversation with Brian Arthur of the Federation of Electronic Industries.

27 *Affordable Recycling: how is it done?*, Executive briefing by the Institute of Grocery Distribution, April 1996, published by Procter & Gamble Limited, p4.

28 Carl Frankel , 'Putting a Premium on the Environment', *Tomorrow*, vol 6 no 3 , June 1996.

29 *Risky Business: How to Cool Global Warming? Show Industry What it Stands to Lose*, Rocky Mountain Institute Newsletter, vol X11, no 1, Spring 1996.

30 Jeremy Leggett, *Climate Change and the Financial Sector: The Emerging Threats, Solar Solutions*, published by Gerling Akademie Verlag, 1996.

31 The International Sustainable Development Research Network, c/o Centre for Corporate Environmental Management, School of Business, University of Huddersfield, Queensgate, Huddersfield, HD1 3DH, UK.

32 Environmental Management Journals, ERP Marketing Department, John Wiley & Sons Ltd, Baffins Lane, Chichester, West Sussex PO19 1UD, UK. The journal titles are *European Environment, Business Strategy and the Environment, Eco-Management and Auditing, Sustainable Development and Business* and *Environment Abstracts*.

NOTES TO CHAPTER 7

1 *Industrial Symbiosis* brochure and *Asnaes Power Station* brochure, available from Asnaes *Power Station* brochure, SK Power Company, Denmark, DK–4400 Kalundborg, Denmark. Phone +45 53 51 15 00.

2 Robert Rasmussen, Industrial Development Council Kalundborg Region – The Symbiosis, P O Box 25, Casa Danica Center, Hareskovvej, 4400 Kalundborg, Denmark.

3 *Theory of Recycling Networks*, Erich J Schwarz, Institute of Innovation Management, Graz, Austria, working paper October 1996.

4 Erich Schwarz & Karl W Steininger, *The Industrial Recycling Network: Enhancing Regional Development*, Research Memorandum no 9501, April 1995, Institute of Innovation Management, Karl-Franzens University of Graz, Johann-Fux Gasses 36, A–8010, Graz, Austria.

5 Information for this section was provided by Ernie Lowe, Indigo Development, 6423 Oakwood Drive, Oakland, CA 94611, USA.

6 This summary is based on data provide by Ernie Lowe, Indigo Development – see note 5.

7 Donella Meadows, 'A City of the Future', *Resurgence*, no 168, January/February 1996. See also Paul Hawken, *The Ecology of Commerce*, Weidenfeld & Nicolson, London, 1993, p213–214.

8 Jon Rocha, 'Let them eat cake', *The Guardian*, 5 June 1996.

9 Michael J Kinsley, *Economic Renewal Guide: How to develop a sustainable economy through community collaboration*, Rocky Mountain Institute, 1994. This section is based on this Guide with some adaptation for UK circumstances.

10 Peace Child International publications include *Rescue Mission Planet Earth, Agenda 21: A Mission Made Possible*, and *Indicators for Action*, all available from:

Peace Child International Centre, The White House, Buntingford, SG9 9AH, UK. *Indicators for Action* is a practical toolkit for children to assess the progress of Agenda 21 in their own neighbourhood, aided by the excellent papers on indicators and *Action Times*, their newsletter. They plan to publish the results of their findings and make a television programme.

11 *Local Agenda 21 Survey 1996*, by Ben Tuxworth and Elwyn Thomas, Environmental Resources and Information Centre, University of Westminster, published by Local Government Management Board, Arndale House, The Arndale Centre, Luton, BEDS LU1 2TS.

12 *Croydon's Local Agenda 21: Living Today with Tomorrow in Mind*, Consultation Draft, Barbara Wilcox, Project, Local Agenda 21, Environmental Health, Croydon Council, Taberner House, Park Lane, Croydon CR9 3BT, UK.

13 *Developing Sustainable Communities: A Field Workers' Manual Part 1*, available from Sue Chapman, Devon County Council, County Hall, Topsham Road, Exeter, Devon EX2 4QU. The document is a joint production involving The Community Council of Devon, Designed Visions of Oxford and Devon County Council. Part 2 will follow, covering design methods and tools and case studies of projects in Devon.

14 Bill Mollison, *Introduction to Permaculture: A Designer's Manual*, Tagari Publications, Australia, 1990/91. Permaculture combines the ideas permanent agriculture and permanent culture to provide a system for creating sustainable human environments.

15 The Environment at Work, developed by Colin Hutchinson and marketed by Chartwell-Bratt Ltd, Old Orchard, Bickley Road, Bromley, Kent BR1 2NE.

16 *Ideas into Action for Local Agenda 21*, published by The Countryside Commission, English Heritage and English Nature, available from Telelink Limited, P O Box 100, Fareham, Hampshire PO14 2SX.

Notes to Chapter 8

1 Chris Fay, Chairman of Shell UK Ltd speaking at a Shell sponsored HSE Forum 25 November 1996, *Business Must Listen and Respond to Earn Public Trust*, available from Shell UK Limited, Shell-Mex House, Strand, London WC2R ODX, UK.

2 It may be more accurate to say that they were re-inspired because they have given a lot of attention to environmental matters for many years.

3 The Environment Council, 21 Elizabeth Street, London, SW1W 9RP.

4 For example: Dr Chris Fay, *Committed to Improvement – How Shell UK approaches the environment* and John Jennings, Managing Director of Shell International, *Getting past sound bites – science, industry and the environment*. Both available from Shell UK – see note 1.

5   Electrolux, Environmental Annual Report 1995, p6.

6   An earlier version of this diagram is in *Vitality and Renewal: A Manager's Guide for the 21st Century,* by Colin Hutchinson, Adamantine, London, 1995, p105.

7   Two excellent books are: Victor Papanek, *The Green Imperative: Ecology and Ethics in Design and Architecture,* published by Thames and Hudson, London, 1995, and Dorothy Mackenzie, *Green Design,* 1991, published by Laurence King, London. A new edition of the latter is expected in 1997.

8   Suzanne Pollack, *Learning to Change: Implementing Corporate Environmental Policy in the Rover Group,* contained in *Greening People,* Walter Wehrmeyer (ed), published by Greenleaf Publishing, Sheffield, 1996, p326.

9   *Integrated Solid Waste Management: a Life Cyle Inventory,* Blackie, 1995, has gone into a third printing of 1000 copies and is available complete with a computer model at £79. The spreadsheet on packaging life cycles is now in a more user-friendly format and is widely distributed in Europe and the USA, including 900 copies to local authorities. The packaging spreadsheet is available from P&G Environmental Affairs Department, Procter and Gamble Limited, PO Box IEE, Gosorth, Newcastle-upon-Tyne NE99 IEE, UK.

10 See for example, Peter Jones 'Total Quality Environmental Management and Human Resource Management', contained in *Greening People,* Walter Wehrmeyer (ed), published by Greenleaf Publications, London, 1996, pp35–48.

11 An approach to exploring personal values is outlined in *Vitality and Renewal,* by Colin Hutchinson, Adamantine, London, p203.

12 The Body Shop report is available on the Internet. IBM's report *Consulting the Stakeholder,* is available from Communications Department, IBM UK, 76 Upper Ground, London SEI 9PZ.

13 Details available from BSI, 381 Chiswick High Road, London W4 4AL.

14 Suzanne Pollack, 'Learning to Change: Implementing Corporate Environmental Policy' in the Rover Car Group, contained in *Greening People,* op cit.

15 Making a Statement, by Andrea Spencer-Cooke and John Elkington, *Tomorrow,* vol VI, no 4, pp 52–57.

16 Suzanne Pollack, 'Learning to Change, Implementing Corporate Environmental Policy in the Rover Group', contained in *Greening People,* op cit.

17 Douglas McGregor, *The Human Side of Enterprise,* McGraw-Hill, New York, 1960. Theory X managers believe that people need be directed, are generally unwilling to accept responsibility and are often lazy. Theory Y managers assume that people behave responsibly when treated as responsible, are capable of self-management and willing to exert effort in the pursuit of objectives to which they are committed.

18 Tim Hart, 'The Role of the Personnel Practitioner in Faciltating Environmental Responsibilty in Work Organisations', contained in *Greening People,* op cit.

Notes to Chapter 9

1   *This Common Inheritance: Britain's Environmental Strategy*, HMSO Publications Centre, P O Box 276, London SW8 5DT.

2   See for example the Advisory Committee on Business and the Environment (ACBE), *Sixth Progress Report to and Response from The President of the Board of Trade and the Secretary of State for the Environment*, published by the Department of Trade and Industry, 151 Buckingham Palace Road, London SW1W 9SS

3   *Indicators of Sustainable Development for the United Kingdom*, by the Department of the Environment, available from HMSO Publications Centre, P O Box 276, London SW8 5DT.

4   See for example, Michael Jacobs, *The Politics of the Real World*, published by Earthscan, London, 1996. Pages 68–9 shows a dramatic widening of the gap between rich and poor in the UK. Worldwide, only New Zealand exceeds the UK in terms of rising income inequality, whereas Italy, Finland, Denmark and Canada have reduced income inequality in recent years.

5   *The New Environmental Agenda: Prospects for Sustainable Development*, by The Rt Hon Tony Blair MP, The ERM Environment Forum, 27 February 1996, available from The Green Alliance, 49 Wellington Street, London WC2E 7BN.

6   Richard Welford and David Jones, *Measures of Sustainability in Business*, Centre for Corporate Environmental Management, School of Business, University of Huddersfield, Queensgate, Huddersfield HD1 3DH.

7   Peter James and Martin Bennett, *Environment-related Performance Measurement in Business: From Emissions to Profit and Sustainability*, available from Ashridge Management College, Berkhamsted, HERTS HP4 1NS, 1994.

8   Fritjof Capra, *The Tao of Physics*, Shambala, Boston, 1975, third update 1991.

9   Fritjof Capra, David Steindl-Rast with Thomas Matus, *Belonging to the Universe: New Thinking about God and Nature*, Penguin, Harmondsworth, 1992.

10 Fritjof Capra, *Web of Life*, published by Harper Collins, London, October 1996.

11 *Accounting for Change: Indicators for sustainable development*, by Alex Macgillivray and Simon Zadek, The New Economics Foundation, 1st Floor, Vine Court, 112–116 Whitechapel Road, London E1 1JE.

12 *Consulting the Stakeholder*, IBM UK Report, 1995.

13 *Traidcraft plc, Social Accounts 1994–1995*, available from Traidcraft plc, Kingsway, Gateshead, Tyne and Wear, NE11 0NE.

14 Richard Evans, Director of Social Accounting, Traidcraft Exchange, Kingsway, Gateshead, Tyne and Wear NE11 0NE. He is willing to respond to inquiries.

15 New Economics Foundation, 1st Floor, Vine Court, 112–116 Whitechapel Road, London E1 1JE.

16 Richard Welford and David Jones, *Measures of Sustainability in Business*, see note 6.

17 The Public Environmental Reporting Initiative (PERI Guidelines). The guidelines were devised by representatives from several large companies, namely AMOCO, BP, Dow, DuPont, IBM Corporation, Northern Telecom, Philips Petroleum, Polaroid, Rockwell and United Technologies. All these have points of contact in the USA and all except United Utilities can be contacted in Europe.

18 UK Environmental Reporting Survey 1994, KPMG, 8 Salisbury Square, London EC4Y 8BB. *Tomorrow*, no 1, vol V, January–March 1995; no 1, vol VI, January–February 1996; no 2, vol VI, March–April 1996; no 3, vol VI, May–June 1996; no 4, vol VI, July–August 1996, Tomorrow Publishing, Halsingegatan 9, SE–113 23 Stockholm, Sweden.

19 Based on the views of John Elkington and Andrea Spencer-Cooke, 'Is there anybody out there?' *Tomorrow*, no 2 vol VI, op cit, p57.

NOTES TO CHAPTER 10

1 In the introduction the books by Collins and Porras, *Built to Last*, Century, London, 1994; Kotter and Heskett, *Corporate Culture and Performance*; Prahalad and Hamel, *Competing for the Future*, Harvard Business Review, July/August 1994, were cited as good examples of distilled wisdom for business success.

2 Dr Howarth from Smith & Nephew was speaking at the IBM launch of their *Consulting the Stakeholder* launch on 3 October, 1996.

3 Ontario Hydro, *Sustainable Development Report for 1995*, copies available from Ontario Hydro, Toronto, Canada, Tel: (416) 592 8986.

4 Martin Leith, *The CLGI Guide to Large Group Events*, 1996, The Centre for Large Group Interventions, Nassaukade 58, 1052 CP Amsterdam, The Netherlands.

5 *Tomorrow's Company Report*, published by the RSA, 8 John Adam Street, London, WC2N 6EZ, 1996.

6 This point is discussed thoroughly by Charles Handy in *The Empty Raincoat*, Hutchinson, London, 1994, Ch 3.

7 See Appendix 2 for a summary of the Groundwork survey, *Small Firms and the Environment*.

8 The same will be true of other countries but it is not easy to obtain information.

9 Electrolux: The Global Appliance Company, *Environmental Annual Report 1996*, available from Group Staff, Environmental Affairs, AB Electrolux, 105 45 Stockholm, Sweden.

10 Warren Bennis, Jagdish Parikh and Ronnie Lessem, *Beyond Leadership: Balancing Economics, Ethics and Ecology*, Blackwell, Oxford, 1994.

11 Peter Block, *Stewardship*, Berett-Koehler, San Fransisco, 1983; Ernest Callenbach, Fritjof Capra, Leonore Goldman, Rudiger Lutz and Sandra Marburg, *Eco-Management: The Elmwood Guide to Ecological Auditing and Sustainable Business*, Berrett-Koehler, San Fransisco, 1983; James C Collins and Jerry I Porras, *Built to Last: Successful Habits of Visionary Companies*, Century, London, 1994; Paul Hawken, *The Ecology of Commerce: How Business Can Save the Planet*, Weidenfield and Nicolson, London, 1993.

NOTES TO APPENDIX I

1 James C Collins and Jerry I Porras, *Built to Last*, Century, London, 1994, p47.

2 Verbal communication with Pierre van Coppernolle, Director of Environment, Rank Xerox, Europe.

3 Jan-Olaf Willums and Ulrich Goluke, *From Ideas to Action*, ICC, 1992, p116 and Frances Cairncross, *Green Inc*, Earthscan, London, 1995, p187.

4 John Speirs, MD of Norsk Hydro UK has often spoken about their environmental work on public platforms. Further information is contained in Willums and Goluke, op cit, p127 and Cairncross, op cit, p203.

5 Stephan Schmidheiny, *Changing Course*, MIT, 1992 p221. See also Willums & Goluke op cit, p89 and Cairncross, op cit, p205.

6 Kim Loughran interview with Alan Knight, 'The Green Knight', *Tomorrow*, no 4, vol V, October–December 1995.

7 David Wheeler, 'Using effective communications to improve environmental performance', in Bernard Taylor, Colin Hutchinson, Suzanne Pollack and Richard Tapper, *The Environmental Management Handbook*, Pitman, London, 1994.

8 Verbal communication with Brian Whitaker, Environmental Affairs Manager, IBM UK and 'Consulting the Stakeholder', report prepared by ECOTEC, who carried out the research.

9 Electrolux and the Environment 1994, *Vision, policy and steps taken*.

10 Joseph J Romm, *Lean and Clean*, Kodansha International, New York, Tokyo and London, Preface pxv.

11 Joseph J Romm, *Lean and Clean*, op cit, p 31–33.

12 Julie Hill, Ingrid Marshall and Catherine Priddey, *Benefiting Business and the Environment: Case Studies of Cost Savings and New Opportunities from Environmental Initiatives*, Institute of Business Ethics, 12 Palace Street, London, SW1E 5JA, p41.

13 *$100,000 Bills on the Shop Floor*, Rocky Mountain Institute Newsletter, vol XI no 3, Fall/Winter 1995, p5.

14 Julie Hill, op cit, p38.

15 Verbal information from Steve Tomlin and Nick Hackett, Reclamation Services Ltd., Catbrain Quarry, Painswick Beacon, GL6 6SU, UK

16 'Unilever in Fight to Save Global Fisheries', *Financial Times* 22 February 1996 and WWF Endangered Seas Campaign literature.

17 Peter White *Engineering for Sustainable Development*, edited by James McQuaid, The Royal Academy of Engineering, December 1995.

18 *The Role of Technology in Environmentally Sustainable Development*, available from The Royal Academy of Engineering, 29 Great Peter Street, Westminster, London SW1P 3LW, 1996.

19 See for example *The Planet Saver's Guide to Holidays* published by NPI, 55 Calverley Road, Tunbridge Wells, Kent, TN1 2UE, 1996.

20 *An Independent Guide to Ethical and Environmental Investment Funds*, Sixth Edition, Holden Meehan, 11th Floor, Clifton Heights, Triangle west, Bristol BS8 1EJ.

21 *Affordable Recycling: How is it done?*, published by Procter & Gamble Limited, 1991.

22 Verbal information from Dr George Howarth, European Environmental Affairs Director, Smith & Nephew Europe Ltd.

23 Robert Worcester, 'Business and the Environment: In the Aftermath of Brent Spar and BSE', lecture given to HRH Prince of Wales's Business & The Environment Programme, University of Cambridge, September 1996. See also The Environment Council, *News*, October 1996 and on Internet http://www.shellexpro.brentspar. com

NOTES TO APPENDIX 2

1 *Small Firms and the Environment*, Groundwork, 85–87 Cornwall Street, Birmingham B3 3BY, 1995.

INDEX

OTHER BOOKS ON BUSINESS AND THE ENVIRONMENT AVAILABLE FROM EARTHSCAN

Factor Four
Doubling Wealth, Halving Resource Use
Ernst von Weizsäcker, Amory B Lovins and L Hunter Lovins

In 1971, the publication of *The Limits to Growth* prompted a ferocious international debate centred on the prediction of imminent resource depletion that it contained. Twenty-five years later, another group of eminent scientists has devised the most radical solution yet to one of the Earth's most urgent dilemmas. In this new Report to the Club of Rome, the authors describe how the value of wealth extracted from the unit of natural resources can grow by a factor of four.

Since the Industrial Revolution, resources have been substituted for labour in order to increase productivity. Today, our overuse of the fundamental resources – air, water, soil, energy and materials – threatens to overwhelm the planet's living systems. *Factor Four* argues that correcting this imbalance need not involve great expense; nor will human progress be halted as a result. Technological and economic progress, in both North and South, can be sustainably managed if we substitute the new index of resource productivity for the old one of labour productivity.

This vibrant and readable text, already a major best seller in Germany, describes an 'efficiency revolution' that may determine our future. Alongside a clear-sighted analysis of the false incentives of the market economy, it presents 50 examples which illustrate how the priniciples of resource productivity are easily implemented, and the benefits readily available. As with all revolutions in technological capability, the pioneers of this approach can be expected to reap considerable rewards; and for the first time, developing countries – who have yet to fully implement the wasteful practices of the major industrial nations – are not starting at a disadvantage.

£15.99 hardback ISBN 1 85383 407 6 352pp colour plates

A–Z of Corporate Environmental Management
Kit Sadgrove

Is aluminium bad for you? What is an environmental management system? Is there an effective substitute for chlorine bleach? *The A–Z of Corporate Environmental Management* provides answers to these and many other questions, and is an invaluable guide to managing a company's environmental impacts.

This practical directory assesses hundreds of products in common use, from aerosols to zinc, via formaldahyde and phosphates. Using an easy-to-understand format, it explains each product's use, its benefits and its environmental risks; recommends safer alternative choices where available; explains issues such as animal testing, eco-labels and recycling; and examines the main impacts of major industries, from aerospace to zoos. Its compact, jargon-free definitions will enable you to produce safer products and communicate your needs more effectively to suppliers. Making extensive use of figures and cross-referencing, this book is ideal for managers who are introducing corporate environmental programmes and risk assessments, and for anyone who needs an objective view of environmental issues in business.

£18.95 paperback ISBN 1 85383 330 4 352pp

THE GREEN OFFICE MANUAL
A Guide to Responsible Practice
Wastebusters Ltd

Businesses are becoming increasingly aware of the benefits of greener working practices, including: waste reduction; energy efficiency; reduced costs; and legal compliance. This highly-accessible, jargon-free guide, written with busy office managers in mind, offers clear, reliable information on environmental issues directly affecting business today. Easy-to-follow advice is accompanied by a range of best practice case studies from industry, education and local authorities. Detailed sources of information are provided at the end of each chapter.

Issues covered include:
* Getting Started;
* Office Waste;
* Purchasing Products and Services;
* Building and Energy Management;
* Transport;
* Environmental Awareness; and
* Environmental Management.

Suitable for businesses of all sizes, schools and educational institutions, and goverment bodies, *The Green Office Manual* is the best available independent guide to greening your office and minimising your costs.

£39.95 paperback ISBN 1 85383 447 5 288pp

CORPORATE ENVIRONMENTAL MANAGEMENT 2
Culture and Organisations
Edited by Richard Welford

No technique or technology can be successful without the human element: the cooperation of everyone involved in the organisation. In this sequel to *Corporate Environmental Management: Systems and Strategies.* Richard Welford and the contributors to this book explore the various organisational and cultural concepts which firmly place the corporate environmental management agenda within the human dimension.

Part One provides an introduction to organisation theory and organisational behaviour. Part Two constructs a picture of the linkage between environmental problems and organisational issues. Problems, challenges, contradictions and complexities are tackled in Part Three, which looks at pragmatic and practical approaches and examines ways in which proactive cultures can be introduced in business. The role of values and leadership and an overarching agenda for human resource management are also considered.

Thoughful and accessible, and containing contributions from Ralph Meima, Tony Emerson, Minna Halme, John Dodge, David Jones and Romney Tansley, this book will be of interest to: human resource mangers; students in business schools and on environmental studies courses; busy managers within larger firms and SMEs; and individuals interested in environmental issues.

£15.95 paperback ISBN 1 85383 412 2 £35.00 hardback ISBN 1 85383 417 3 224pp

BUILDING CORPORATE ACCOUNTABILITY
Emerging Practice in Social and Ethical Accounting, Auditing and Reporting
Simon Zadek, Peter Pruzan and Richard Evans

'*Building Corporate AccountAbility* is a landmark book in the development of standards for the social auditing of corporations. For the first time it puts between covers specific examples of how companies go about the job of measuring and evaluating the social impacts of their businesses. It should become an indispensible resource not just for socially-conscious activists but for anyone in the business world who is concerned about the responsibilities of a company in the broadest sense.'

MILTON MOSKOWITZ, CO-AUTHOR OF THE 100 BEST COMPANIES TO WORK FOR IN AMERICA

'This publication combines a clear vision of why social and ethical priniciples should be mainstreamed into business practices, with a set of practical tools and case studies to illustrate how. I recommend it to business people, consultants, academics and students who are interested in understanding what will undoubtedly be one of the key strategic issues facing business in the 21st century.'

JANE NELSON, RESEARCH DIRECTOR, PRINCE OF WALES BUSINESS LEADERS FORUM

Social and ethical accounting, auditing and reporting is emerging as a key tool in response to calls for greater transparency in business, and as a means for managing companies in increasingly complex situations where social and environmental issues are significant in securing business success. This is the first book to address these issues and their implications for the development of corporate responsibility. The editors introduce a historic overview of developments and a methodological framework that allows emerging practice worldwide to be analysed, understood and improved upon. This innovative book, which includes nine case studies from world-class businesses, will be of enormous value to business managers and students.

Contents: Introduction • Why Count Social Performance? • How to Do It • Accountable Futures • Sbn Bank, Denmark *Professor Peter Pruzan* • Traidcraft UK *Richard Evans* • The Body Shop International, UK *Maria Sillanpää and David Wheeler* •Ben & Jerry's Homemade Inc, USA *Alan Parker* • Municipality of Aarhus, Denmark *Carl-Johan Skovsgaard and Tom Christensen* • Woyen Molle, Norway *Lise Norgaard* • Consumer Cooperative, Italy *Alessandra Vaccari* • Van City Savings & Credit Union, Canada *Cathy Brisbois* • The Practice of Silent Accounting *Professor Rob Gray*